AF566941

ENERGY SCIENCE, ENGINEERING AND TECHNOLOGY

# ENERGY EFFICIENCY

# WHAT IT IS, WHY IT IS IMPORTANT, AND HOW TO ASSESS IT

# Energy Science, Engineering and Technology

Additional books in this series can be found on Nova's website under the Series tab.

Additional e-books in this series can be found on Nova's website under the e-book tab.

ENERGY SCIENCE, ENGINEERING AND TECHNOLOGY

# ENERGY EFFICIENCY

# WHAT IT IS, WHY IT IS IMPORTANT, AND HOW TO ASSESS IT

XAVIER CHAVANNE, PH.D.

New York

For permission to use material from this book please contact us:
Telephone 631-231-7269; Fax 631-231-8175
Web Site: http://www.novapublishers.com

## NOTICE TO THE READER



## Library of Congress Cataloging-in-Publication Data

Chavanne, Xavier.
Energy efficiency : what it is, why it is important, and how to assess it / Xavier Chavanne.
pages cm
Includes bibliographical references and index.
ISBN 978-1-62808-764-2 (hardcover)
1. Energy conservation. 2. Energy consumption. I. Title.
TJ163.3.C425 2013
333.791'6--dc23

2013029302

*Published by Nova Science Publishers, Inc. † New York*

# Contents

# List of Figures

# List of Tables

# Abstract

**What is Energy Efficiency?** J. Watt and S. Carnot used the amount of coal burnt by steam engines per unit of water lifted times height to compare the efficiency from one design to another, or to an ideal machine. Likewise for any well-defined industrial system - agriculture based ethanol production, phone manufacture, transport... - one can determine from the system inputs and outputs its direct consumptions - electricity, fuels... - per unit of output.

However, the indirect consumptions of auxiliary systems to produce the requirements of the main system, including its equipments, should be taken into account. Furthermore, an energy system like ammonia or agro-ethanol industries can also consume a part of its main resource to fuel its processes.

Different indicators take into account these contributions. They do not provide the same value for a same system and its flows. Is there one more complete than others?

**Why is Energy Efficiency Important?** Energy consumptions, along with other requirements like work hours, steel contained in equipment... per unit output have decreased during the industrial revolution, resulting in abundant, cheap and useful energies, materials and services. Thus, thanks to fundamental and applied knowledge like thermodynamics and engineering sciences, the energy use rate for the ammonia production fell by a factor 5 and that for steam engine by a factor 250. However, the very theories responsible for these progresses predict their limits. Is economic growth still possible within these limits?

These abundant energies permitted the seven fold growth of the population and its well-being since 1800. In turn growth and the search of comfort increased the demand of useful energy. For American car drivers only high prices seem to be able to stop this trend.

Raw energy resources on Earth are extremely abundant; total solar radiation input is equivalent to 10 000 time the energy consumption by humans, and organic carbon in the earth crust could provide for current consumption for almost 10 000 000 years. However, available technology and their efficiency reduce the accessible part or reserves considerably. The choice of processes for the extraction and their efficiency depend on the characteristics of the deposits (concentration, state, depth...).

**How do we Assess Energy Efficiency?** The methodology proposed in this book links the efficiency at the system level to the data - flows and established knowledge - found at the process level. This analysis determines the dependence of the system efficiency on physical characteristics of its processes. Unless this is done these characteristics may be sources of large errors - by factors of one hundred or more -. The suggested methodology saves time of analysis and gives a realistic assessment of the remaining uncertainties.

Complete energy systems cannot dissipate more energy than that they extract, directly or indirectly. Historic exploitation of underground coal could not run a steam engine for operations which requires more coal than it can lift. Can the agro-ethanol industry operate without external energies, *i.e.* it is more than self-reliant?

# Preface

The topic of this book is the energy efficiency in industry, *i.e.* the potential reduction of the energy dissipated along a chain of processes from the natural resources extracted by humans, or primary energy, to the useful forms of energy such as electricity, transport fuels, mechanical energy, heat or even ammonia. It also deals with the energy required to produce consumer goods like phones and capital items like tractors, or to provide services like freight shipments, and information and communication technologies.

The search on better energy efficiency of processes started even before the very notion of energy, or the equivalence of heat and work, was established. Indeed the first improvements on the steam engine to reduce its coal requirement at identical work took place before the foundation of the thermodynamic theory.

The present study does not cover new developments in the field of energy conversion, as was the fundamental understanding of work production from heat in the steam engine (which was at the origin of the thermodynamic theory). Instead, it gathers and analyzes critically established knowledge and data in both fundamental and engineering sciences to determine the efficiency of any industrial system (from coal exploitation to phone manufacturing), and to compare between possible designs for potential gains.

The book aims equally to bridge a growing gap between this specialized and complex information about processes, and questions about energy efficiency at the economy or society level. Consequently it intends to address a large audience from physicists and engineers to economists, executives and teachers, i.e. anyone with an interest in energy.

## What is Energy Efficiency?

The chapter of the book following the introduction shows that, even for a well-defined system and its materials and energy flows, the assessment of energy efficiency is not as simple as it may appear at first.

Within the book the efficiency/inefficiency of the system is defined by the energy required per unit of its resource input or its production output. However, the rate changes with the choice for the denominator and with what are considered energy requirements. Is the energy consumed to manufacture the equipments and materials, or to produce the electricity and fuels used by the system, included in the requirements? Must we count all the energy consumed or only that dissipated in the processes? Are the requirements limited to external primary energy?

Moreover, a practical rate like the energy consumed per unit of joule J of ethanol produced can give a value higher than one. It does not mean that there is any violation of the energy conservation, or the system dissipates more energy than it extracts. A more fundamental rate can be established based on the thermodynamic laws. It leads to the notion of the self-reliant energy system, which uses only the energy resource it extracts and processes to meet all its requirements, both fuel and feedstock. By its mere existence our whole energy system, from petroleum industry to hydroelectric production, is self-reliant. But not each of its components is necessarily self-reliant. The notion is important not only in terms of the fundamental significance of the efficiency of a system, but also for the viability of entire societies and their thriving economies, as the example of an ancient coal exploitation demonstrates.

## Why is Energy Efficiency Important?

Because of the importance of the energy - and subsequently of energy efficiency in terms of energy savings - for the activities of our own society and for our personal life, the book in the third chapter also deals with these connections between industrial efficiencies and the economy or the human well-being.

The energy efficiency issue has gained renewed importance with the recent rise of hydrocarbon prices. France, the author's country, imported roughly 85 G$ - billion dollars - worth of crude oil, refined petroleum products and gas in 2012. At the end of 90s the annual trade bill amounted to a mere 20 G$ (in constant $ value). This rise represents 2 to 3% of the annual France Gross Domestic Product GDP, small as a percentage but with a staggering impact on economic growth and national debt levels. Further improvements in efficiency in the transport sector or in the thermal insulation of buildings could significantly alleviate this financial burden.

The objective of the third chapter is equally to distinguish among the various energy related notions in economic and social themes those that are not relevant to the energy efficiency in its strict definition.

Thus an economic system is judged efficient if it generates a large added value. The latter represents the sales of the system outputs less the cost of its inputs. The added value is distributed in salaries, taxes and surplus capital. In the long run, especially for basic industries such as electricity generation and ammonia industry, the value benefited from the improvements of the manufacturing processes and the lowering of their energy and material consumption rates. Along with rising productivities in labor and capital investments, these gains in the basic industries permitted to decrease the unit price of their products, on which the whole economic activity has drawn to thrive. The price reached such a low level that it reduced the weight of the economic value of these industries in the GDP, less than 10 percent in developed countries.

However, the unit price can equally be driven by the balance, or imbalance, between demand and supply, or, in the short term, by government ill-defined regulations or by speculative markets, as observed in the case of the industry and market of the photovoltaic modules since 2000. Mechanisms of price fixing add another layer of complexity, which may well hide fundamental trends in the efficiency of processes and so in the long-term cost of products.

Thermodynamic theory, which has explained and guided the 250 fold efficiency gains of the steam engine since its inception, also determines the limit to these gains, as seen in the last chapter of the book. For the last decades the extraction and transformation of each barrel of crude oil have required increasing amounts of physical resources - energy, equipment...-, despite evident progress in the specific efficiencies of the processes involved. Because of the oil high demand - except during crises - due to its essential function in running nearly all economic activities, its prices have definitely followed the physical trend. Only the thorough technical study of extraction processes and their improvements as well as that of the characteristics of the deposits to be developed to maintain production can reveal future developments. In summary, not only "it's the economy, stupid", but also "it's thermodynamics and engineering, stupid".

At the consumer level energy saving can also be obtained by the decision to drive less, to buy a smaller car, or to reduce the heating temperature in his/her house···, which is in fact independent of efficiency. It is more often than not motivated by the final price of the useful energies like electricity and transport fuel, as the behavior of the typical American drivers proves. They are essential consumer costs which governments try to control to deflect the social pressure.

Moreover, the price paid by the consumer for other goods and services such as cars and insurances has little relationship to the energy costs to produce them. As a result of productivity and efficiency gains of basic industries the energy cost component is generally similar to the one in the country GDP. At the personal level incentives for saving energy are thus reduced.

At first glance the abundance of the energy resources on Earth surface, or close to it, is independent of the efficiency of the processes employed to extract and make them useful. The resource base is actually huge compared with the human demand, either in terms of energy of flows, like solar radiation, or energy stores such as fossil organic carbon in the crust of Earth. However, only a very tiny fraction of this energy resource is amendable to exploitation. Its characteristics - concentration, accessibility, state... - do allow the use of efficient process. Other portions of the resource can require more energy to extract and transform them than they actually contain, making their extraction physically non-viable.

## How to Assess Energy Efficiency?

The fourth chapter describes a methodology to assess the rate of energy consumption of a complex system and the dependence of this indicator on a variety of physical and technical quantities.

The methodology rests on the decomposition of the system on its much simpler parts for which a local consumption rate - *a priori* independent of the system one - is defined and calculated from available and accurate information (flow data and established relationships at process level). Thus is isolated the mechanized operations at the farm in an agro-ethanol industry. Its efficiency is expressed with the volume of diesel required per unit of cultivated surface. From the local rate of the operation is determined the operation contribution to the global rate of consumption. The relation between both rates provides some of the variables of the system efficiency. The conversion from the consumption of diesel per ha of crop acreage to the consumption per joule of ethanol requires the yield of crops per hectare, the mass of sugar or starch in the harvest and the fraction of sugar converted into ethanol. The physical analyses of these variables and others from different operations define possible future gains or limitations on the final efficiency of the agro-ethanol industry.

The fifth and last chapter dwells on the self-sufficiency of energy systems, or in other words, whether systems can only rely on the resource they extract for their operations. Coal extraction from underground mines necessitated efficient steam engines to lift the coal and pump the excess water. These engines had to consume only a fraction of the coal lifted at the surface for coal exploitation to be viable and self-reliant.

Agro-ethanol industry is a more complex energy system as it requires the outputs of other energy systems - diesel, electricity, ammoniac... -, and consequently depends on their efficiency. From the industry actual direct consumptions (electricity, fuels...) and by replacing existing processes to produce internally these requirements, the system can be made self-reliant, consuming its own production and/or resource. Hence its real efficiency can be determined. It may well be possible that the industry does not produce enough energy to fuel its own processes.

**And More**

The reader is invited to learn more about the agro-ethanol industry and its efficiency throughout the book. Other examples like steam engines, the ammonia industry, phone manufacture... are also used to illustrate and support the various arguments developed in this book about energy efficiency.

Feedbacks on this study are welcome. Oversights on such a complex matter cannot be ruled out.

The author thanks the prof. J.-P. Frangi, director of the Institut Universitaire Professionnel Génie de l'Environnement at the university of Paris Diderot for his help and remarks. Without him this work would not have been possible.

He is also grateful to P. Brocorens, P. Alba and B. Durand for their thorough review. P. Brocorens is a young researcher on advanced chemistry at the University of Mons working on the properties of new materials like conjugated polymers and composites. P. Alba is a retired engineer of Elf Aquitaine where among various positions he had been in charge of the economic studies of the company. B. Durand is an expert of petroleum geochemistry and had been responsible of the geology-geochemistry department at Institut Français du Pétrole (IFP). He had been also director of Ecole Nationale Supérieure de Géologie in Nancy.

This book also benefited from the review of J. Heissner, promoter of a novel project of a self-sustainable village in New-Zealand. I wish him success in his project.

*X. Chavanne*

Physicist (PhD, member of Société Française de Physique) and research engineer. He worked on various topics in academic and applied areas: study of the turbulent natural convection both experimentally and theoretically, modeling of the convection in glass furnaces at St Gobain company (with physical models and theory), development of soil moisture sensors for both fundamental and industrial objectives (use of a dielectric principle). Since 2004 study of the efficiencies of different industrial systems in areas from the telecommunications to basic industries and to transport.

September 2013

# Chapter 1

# Introduction

## 1.1. First Mentions of Energy Efficiency

The first evocations of the efficiency of an industrial system in modern terms can be associated with the improvements of the steam engine by J. Watt, and with the work of S. Carnot to establish a theory of fire machines at the turn of the nineteenth century. Both J. Watt and S. Carnot used as an indicator of efficiency the quantity of coal burnt, or its calories, per unit of the product of mass of water lifted by the height. Indeed, the notion of energy itself was introduced later by the experiments of J. Joule and the establishment of the thermodynamic theory by R. Clausius and W. Thomson.

The steam engine was first developed successfully in Great Britain in 1712 by T. Newcomen to access water filled coal mines [1,2]. Because of the wood disappearance and the easy access at the surface or shallow depths of peat and coal in some regions of Europe (Holland, Great Britain, Wallonie region in Belgium, the basin of la Loire in France...), the fossil resources were exploited as early as the thirteenth century. Due to a rising demand the production had to be increased.

The Newcomen machine had a very low yield in terms of conversion of coal heat into work (less than 0.3% as established later; see the details in the chapter 5). Successors of T. Newcomen managed to increase this efficiency, most notably J. Watt after 1769 (separate condenser among various modifications). One of Watt's sale pitches was that his new machine was able to save 75% of coal by comparison with the previous design [3]. The efficiency was all the more important as J. Watt's improvements and modifications allow the use of the steam engine in other industries (textile, iron...) to provide mechanical energy. Thanks to Watt and other inventors the steam engine became more efficient - still less than 5% in 1824 - while more available and reliable to produce energy than other means such as muscular work, windmill or even water mill. After 1830 with the locomotive the steam engine was used in the transportation of freight and people.

Around 1820 the aim of the physicist S. Carnot was to determine the ideal transformations along a cycle in a "fire" machine (not only a steam engine) to produce a maximum of work from a fixed amount of heat [4]. It was equivalent to solving an efficiency problem with the deduction of the fundamental factors, chiefly the temperature difference between the hot and cold sources rather than the level of gas pressure as believed at that time, to be most important (at least for an ideal cycle). His study founded the theory of thermodynamics with notions like thermodynamic cycles, reversible transformations, adiabatic and isothermal compressions or dilatations of a perfect gas... Thanks to his work large and fast improvements were made possible with the existing machines (the steam engine benefited of the theory as shown in the chapter 5), and new machines were conceived (internal combustion engine, gas turbine). Some modern steam engines in power plants can reach a yield of 45% and gas fired plants can surpass 55% owing to two combined cycles.

## 1.2. Structure of the Present Book

In the early industrial revolution energy efficiency of processes was one of the major preoccupations of engineers to save expensive coal or to increase the outputs of an industry with the same amount of coal. Energy efficiency has always been a priority when the supply of energy is restricted resulting in high prices, as observed in the beginning of 19th century for coal and nowadays for crude oil. Obviously, as it was also the case in the nineteenth century, energy efficiency is not the only factor to consider when developing a process or a system at the industrial scale. They must also fulfill economic or environmental constraints such as labor and equipment productivity, which can further impact the energy consumptions.

The purpose of the book is not to present new developments on energy conversion, as was the understanding of work production from heat in the steam engine. But rather it will show how to gather and analyze critically already established knowledge and data of both fundamental and engineering sciences to determine the efficiency of a large range of industrial systems.

It also deals with the role of the energy efficiency in the economy and human societies.

The book in the second chapter establishes first a definition of the efficiency for any system. It is expressed with an indicator based on the rate of consumption of the system, *i.e.* its energy consumed per unit of its output. It turns out that different indicators for a system and its flows are possible. A practical rate is estimated from the direct inputs and outputs of the industry, like coal and lift work in the case of the steam engine. Its scope can be enlarged to include indirect consumptions like those to extract and clean the coal used by the steam engine. However, in the case of energy systems like steam engine or coal exploitation, the application of the energy conservation suggests a different and more fundamental rate based in the energy dissipated.

The third chapter shows how the efficiency gains and its limitations still play an important role in our societies, its economy, human well-being and comfort, and our extraction of natural resources although it is not as apparent as it was in the times of Watt and Carnot due to the very progresses of efficiency since then. Other factors have also to be considered like the cost of production in monetary value, the prices of energies and products, energy conservation by end consumers, as well as the abundance and characteristics of the energy resources.

The fourth chapter is dedicated to a methodology and its mathematical tools to derive a rate for the system under study and the dependences of this rate. Energy efficiency can be studied for quite simple systems like the steam engine as well as for more complex ones such as the production of ethanol from agriculture - agro-ethanol - or the phone manufacture. The main principles of the analysis are the decomposition of the system towards simpler operations of few processes or even just one, and the determination of the natural rates of consumption of these operations. These local rates, deduced only from data at operation level, are much less variable than the rate of the whole system. A reverse process allows to derive from the local rates the latter rate. As a result of the process, the physical and technical variables on which the system rate depends are identified. In the process, and thanks to different sources and/or established knowledge in sciences and technologies, large errors on data and on assumptions are tracked. The suggested methodology also provides tools to propagate the remaining uncertainties from the raw data to the system rate. This allows to quantify the compromise between the accuracy on the rates and the time required for data gathering and analysis.

The last chapter deals more specifically with the energy systems. Among them are examined self-reliant systems. The resource they use must provide both the fuel consumed in their processes and their feedstock. Historic industries such as the exploitation of a coal basin offer examples of autonomy due to the necessity themselves to use local resources for production. Significantly, in industrial times only fossils fuels show their ability to approach self-sufficiency. If they are to be substituted for conventional and dominant systems, alternatives systems like the agro-ethanol production must also demonstrate this capability. This can be achieved thanks to modifications of present systems with existing processes. However, the resulting rate of dissipation expressed as the energy dissipated per unit of total input of the system must be lower than one in order for the self-reliant system to produce a surplus output.

**Note about Unit Symbols** Due to the difficulty to define an equivalence between the different forms of energy such as between final and primary energies, the unit of each form of energy is accompanied by a subscript informing about the form.

Thus the subscript **LHV** in $J_{LHV}$ stands for the low heat value of a fuel, **HHV** in $J_{HHV}$ for its high heat value, **e** in $J_e$ for electricity, **m** in $J_m$ for the mechanical energy, **th** in $J_{th}$ for the thermal energy, **ph** in $J_{ph}$ for the energy of photons, **EF** in $J_{EF}$ for the final energy - part of the direct requirements $E_D$ - consumed by the system under study, **EP** in $J_{EP}$ for the overall primary energy $E_{EP}$ used to produce $E_D$ by the system *aux*...

The subscripts of units are also used for other quantities if necessary. For instance the volume of water lifted over a fixed height is measured by $m_w^3.m_h$.

The use of subscripts for the symbol of units is not recommended by the Comité International des Poids et Mesures (CIPM) in charge of defining the Système international d'unités (SI) [52]. All energies have a unique (derived) SI unit, joule, based on their equivalence in terms of heat value. But, as we have seen, the different forms are not equivalent in terms of usefulness for the consumer. Otherwise the notion of efficiency itself can be questioned.

Notations like J of electricity or J of primary energy could be used. But for concision we prefer to use the convention introduced above. We follow the rest of the norms enacted by the CIPM.

# Chapter 2

# A General Definition of Energy Efficiency?

The chapter presents the various indicators encountered in the literature to define the energy efficiency of an industrial system. All of them can be presented in the form of an energy consumption or dissipation per unit of the service performed by the system. From the diagram of materials and energy flows at the system boundary and within it, the definition of each of theses rates is specified, as well as the relations between them. The rates range from the practical one, readily available from the raw data of flows of the system but of limited scope, to a more complex one taking into account, however, the thermodynamic laws.

The range of choice is especially large in the case of energy systems, of which feedstock is a raw form of energy and output is a more useful one for human societies.

Among the energy systems we can define the self-reliant ones, of which unique input provides both fuel and feedstock to the system. Such systems existed in historic times by necessity. Our whole energy system is fundamentally a self-reliant one, demonstrating its relevance. The definition of the rate of self-reliant systems is quite straightforward.

Actual industrial systems and their processes - the ones studied along this book - must fulfill constraints other than the energy efficiency (economic productivities, environmental regulations...). Each of these constraints can be expressed by a physical rate.

## 2.1. Rates of Energy Consumption of an Industrial System

At the Watt's and Carnot's times the efficiency of a steam engine was measured by the quantity of coal required to lift 1 $m^3$ of water over a height of 1 m (with the units used by S. Carnot in his book [4]). The rate was derived directly from the values of the engine input and output.

We generalize this practical definition to any industrial system from the agro-ethanol (ethanol from agriculture products) production to the phone manufacture, the freight transport or the ammonia manufacture.

### 2.1.1. The System and Its Flows of Energy and Materials

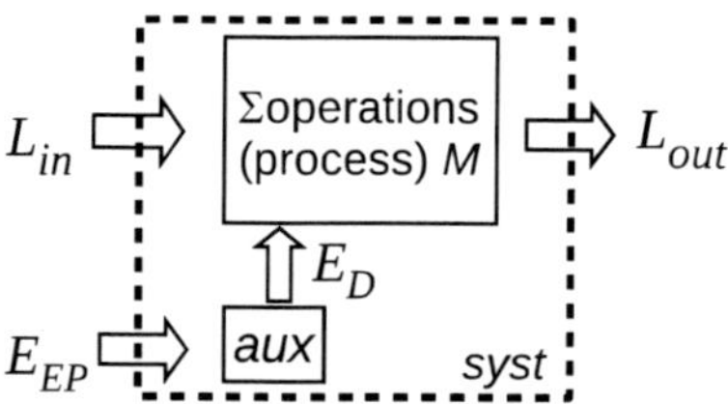


Figure 2.1.1. Schematic diagram of materials and energy flows for an industrial system *syst*. The main operations *M* provide from the feedstock $L_{in}$ of raw material or services the output $L_{out}$ of products. To achieve this work they consume directly the flow $E_D$ of energy and materials. Auxiliary operations *aux* have transformed the flow $E_{EP}$ of primary energy into $E_D$.

The system *syst* is modeled by the balance of its materials and energy flows over a fixed period such as a year or an hour (Fig. 2.1.1). Its boundary is fixed by its inputs, feedstock $L_{in}$ and energy consumption $E_{EP}$, and its output $L_{out}$. The latter can be the volume of the agro-ethanol production, the number of phones manufactured, the number of tonne.km of fright shipments or the mass of ammonia $NH_3$. To achieve this work, during the same time the system consumes some natural resources $E_{EP}$ extracted mainly for their heat value - crude oil, gas, coal, heat from nucleus fission, hydroelectricity... - i.e. primary energy (see their definition in the appendix).

Within the whole system *syst* we distinguish the main operations $M$, which actually perform the production, from the auxiliary operations *aux*, which provide the direct requirement $E_D$ of $M$ from the primary energy $E_{EP}$.

Along the book an operation represents the aim (ethanol dehydratation, transport...), while the process is the means by which the operation is performed (ternary distillation or use of molecular sieves, transport by heavy duty truck or by train...). The values of the consumption rates depend, among other factors, on the choice of the process.

The consumption $E_D$ represents usually the data directly available at the process level of the main operations $M$, either final energies such as petroleum products, processed gas and electricity, or direct energies such as mechanical work, cold and heat flows. The flow $E_D$ includes as well materials of equipment and chemicals since their manufacturing in the operations *aux* requires energy. The balance of requirements for the steam engine should thus comprise not only the coal consumption but also the materials to make the boiler, cylinder, piston, rods, pipes, chokes... of the engine (often a lower contribution over engine lifetime).

**Difference between the Inputs $L_{in}$ and $E_D$ of the Main Operations** The input $L_{in}$ refers first to the feedstock - natural resources or products of an upstream industry - of which transformation by the operations $M$ of the system produces the output $L_{out}$. In case of the agro-ethanol industry the input $L_{in}$ is the flow of the harvested agriculture plants such as corn or sugar cane, while for the phone manufacture it represents the amount of raw materials to obtain the various parts of the phone[1]. In the case of the transport by truck the load to be delivered may be considered the input $L_{in}$. At a coal mine the input is the raw coal extracted to be sorted out and washed before its sale (output $L_{out}$).

The quantity $E_D$ represents the resources in form of energy and materials directly consumed by the operations $M$ of the system. These resources are transformed into energy and materials not anymore usable for the system (heat or cold too close to ambient temperature, wastes, effluents...).

In a ideal system the consumption $E_D$ would tend to zero whereas the flows $L_{in}$ and $L_{out}$ are the ultimate quantities. As a result, the rates such as the direct rate $R_D$ (eq. 2.1.1) would also tend to zero.

[1]This latter industry can be extended to include the extraction operations according to a life cycle analysis; the input of the phone manufacture are then the natural feedstock for its components.

However, the boundary between the flows $L_{in}$ and $E_D$ is not always clear. Actually for the practical and straight determination of the rate $R_D$ from the data, the above rule is not respected.

The plastic items of the phone are mostly made out of petroleum products and so of crude oil. Only oil consumed and definitely lost during the production should be counted in the quantity $E_D$. Is it really the case in the LCA databases?

In the case of the steam engine, all the heat value of the coal burnt in the engine is included in the consumption $E_D$ (as a consequence the input $L_{in}$ is non existent). We should have removed from this consumption the engine output, that is the mechanical energy of lift (which was negligible for the early steam engines). This treatment implies the use of the same energy unit for all flows instead of the units of the raw data.

Similarly the natural gas in an ammonia factory is used as feedstock and fuel but, for convenience, the whole is counted as requirement $E_D$.

Conversely, a part of the feedstock like the bagasse of the sugar cane in the ethanol factories of Brazil can be used as fuel. This part should be included in the flow $E_D$, but the rates of eq. 2.1.1 and 2.1.2 actually omit it.

As we see the distinction between the quantities $L_{in}$ and $E_D$ becomes blurred, especially in the case of energy systems of which flow $L_{in}$ is a form of energy. The section 2.2 studies more thoroughly the case of energy systems.

**Complex Systems** The apparent simplicity of the diagram of Fig. 2.1.1 hides in reality more complex systems in both $M$ and *aux* parts. Even systems such as the steam engine or the ammonia factory, when studied in details (see the section 5.1.1 of the chapter 5 for the first one, and section 3.1.1 of the chapter 3 for the second one), are not as simple as they seem.

A complex system consists of different operations of which consumptions can be added to obtain the overall consumption $E_D$ (or overall consumptions by type of requirements like steel and final energies).

For instance the ethanol industry can be broken down into the farm stage, the plant transport, the factory and the ethanol distribution. Each one can be further decomposed into simpler operations (more in the section 4.3 of the chapter 4).

Importantly, due to this complexity of industrial systems, the value of the rate, once its definition is fixed, varies according to physical and technical quantities. The proposed methodology to derive these quantities and their influence is exposed in the chapter 4.

### 2.1.2. Rate of Direct Consumption $R_D$

From the direct and readily available flows $E_D$ and $L_{out}$ (or $L_{in}$[2]) at the main operations $M$, the efficiency indicator is expressed as:

$$R_D = \frac{E_D}{L_{out}}. \tag{2.1.1}$$

The rate $R_D$, or $R_D^M$ in case of confusion[3], is the consumption of direct energies and materials of the system *syst* per unit of its products.

The indicator $R_D$ offers a tool of comparison for the efficiency readily available between systems with identical forms of consumption $E_D$ and output $L_{out}$ (and usually having an identical resource $L_{in}$). The consumption of coal calories per unit of work or electricity has thus guided the efficiency progresses for the steam engine. Energy performance of an ammonia factory is still measured by the amount of fuel required per tonne of $NH_3$. A comparison between different trucks or other modes of freight transport can be drawn from the direct rate in liter of diesel per one tonne of freight shipped over one km. The efficiency of the heat pumps is measured by the amount of electricity required per unit of heat provided by the pump (the inverse of the coefficient of performance of a heat pump).

In the case of complex systems such as the agro-ethanol industry, their direct requirements are often in various forms like electricity, fertilizers, seeds, gas, diesel... or equipment such as tractors, trucks and reactor tanks, making difficult an aggregate value for the rate $R_D$. It is thus limited in its scope to systems consuming the same type of flow $E_D$ like the steam engine or the ammonia industry. Otherwise, a rate $R_{Dk}$ is defined for each type $k$ of direct consumptions. Thus we can define the rates of consumption of concrete, steel or fuels to built an electrical generator and draw a comparison (like in the paragraph 3.3.2 of the chapter 3).

[2]In some cases the rate $R_D$ is better defined with the system input $L_{in}$ such as in the semiconductor industry, where the surface input of the Si wafer is preferred to define $R_D$, or in petroleum refinery where the crude oil is the reference.

[3]Confusion can arise when, as an intermediate step, the efficiency of a subsystem *subM* of the main system $M$ has to be studied independently with its rate $R_D^{subM}$ (more in the section 4.3.2 of the chapter 4).

### 2.1.3. Rate of Primary Energy Consumption $R_{EP}$

As shown in Fig. 2.1.1, the flow $E_D$ is produced by auxiliary operations *aux*. From the study of each operation we can derive a specific factor $\beta_{aux}$ of conversion of $E_D$ into the quantity of primary energy $E_{EP}$ required. Hence a more general rate is defined for the system *syst*:

$$R = \frac{E_{EP}}{L_{out}}, \qquad \text{with}\, E_{EP} = E_D \cdot \beta_{aux}. \tag{2.1.2}$$

The rate $R$, or $R_{EP}^{syst}$, is the consumption of primary energy of the system *syst* under study per unit of its products.

The factor $\beta_{aux}$ includes all the primary energy used by the operations *aux*, whether as feedstock or fuels. Thus it takes into account the energy consumed to extract the resources from which $E_D$ is obtained, transport and transform them. For instance a coal mine is the auxiliary of a steam engine by supplying its fuel. The coal burnt in engines to extract bituminous coal from an underground mine in la Loire France in 1927, to clean and transport it, amounted to about 3% of its heat value (see the section 5.1.2 of the chapter 5). Hence, as one tonne of the coal has a high heat value HHV of about 29 $GJ_{HHV}$, the factor $\beta_{coal}$ for the mine operating as auxiliary is $\beta_{coal} = 30\ GJ_{EP}.t^{-1}$. In the case of transport, the petroleum industry is the main auxiliary system. The fuels used by the petroleum industry represents from 12 to 20% of the diesel HHV [20]. As one liter of diesel has a high heat value of around 38.5 $MJ_{HHV}$, the factor $\beta_{diesel}$ to convert diesel into primary energy varies from 43 to 46 $MJ_{EP}.l^{-1}$.

The factor $\beta_{aux}$ can include the savings generated by process integration in factories. Heat exchangers provide the requirements of some operations $M$ by recovering the waste heat of other main operations, depending on its temperature level. Heat exchangers are considered processes of the auxiliary operations. By comparison with a system without waste recovery, they reduce the requirement $E_{EP}$ at same flow $E_D$. However, available data for the flow $E_D$ may already integrate this effect.

Moreover, all the contributions to the rate $R$ from the operations $M$ of the system are made homogeneous to a primary energy, which allows their comparison and their addition.

On the other hand some drawbacks of the rate $R$ make important to calculate and keep the rate $R_D$ of eq. 2.1.1.

Firstly, the latter rate permits to separate the study of the efficiency of the operations $M$ from that of its auxiliaries considered as external systems. Hence for a same main system and its consumption $E_D$, other choices of auxiliary operations are possible with different coefficients $\beta_{aux}$ (the forms of primary energy being either the same or different).

Secondly, the search of direct consumptions for $E_D$ allows a better control of potential errors as they are close to the data measured at the process level of the main system $M$ under study.

Thirdly, the rate $R_D$ is useful to study criteria other than energy efficiency such as the costs of equipment per unit of output (see next section).

**Equivalence between the Different Forms of Energy** A last reason to keep the rate $R_D$ is that, even expressed by their heat values, primary energies are not truly equivalent relative to the service they offer at the end of the chain of transformations and transport. Indeed, for a same output of a product in a system, the amount $E_{EP}$ required can vary depending on the primary energy used and the associated processes. One important example is the production of electricity in thermal power plants, even modern ones. The energy consumption varies from 3 $J_{EP}$ of primary energy for 1 $J_e$ of electricity in a nuclear plant, to less than 2 $J_{HHV}$ for 1 $J_e$ in a combined cycle gas plant. This problem of equivalence stems from the fact that our societies and economies definitely require end use energy, like the electricity at the socket, and not primary energy. It shows the importance of the energy efficiency of a system to define a sort of equivalence between primary energies.

Furthermore the abundance of a primary energy should also be factored. Even expressed in terms of electricity at the socket - and thus taking into account the efficiency of the power industries from the extraction operation -, uranium is more abundant than crude oil or natural gas. The section 3.3 of the chapter 3 provides some details about this point.

**Relationship between the Factor $\beta_{syst}$ of a System *syst* and Its Rate $R^{syst}$** Ancient underground coal mines operated with steam engines used as auxiliary system to provide the work to lift water and coal (see the example of la Loire basin in 1927 in the section 5.1.2 of the chapter 5).

From this example one can wonder what is the relationship between the quantity $\beta_{aux}$ - conversion of mechanical lift into the coal consumed by the engines - and the rate $R^{aux}$ of the auxiliary system itself - the steam engine -.

Indeed the auxiliary system and its efficiency can be studied independently of the main system $M$.

Let's examine a system *syst* used as auxiliary to supply the direct requirements $E_D^M$ of a main system $M$ from an overall consumption $E_{EP}^M$ of primary energy. On the other hand, the system *syst* is viewed as an independent one with its main input $L_{in}^{syst}$ and output $L_{out}^{syst}$. It requires the consumption $E_{EP}^{syst}$ of primary energy (*syst* includes its own auxiliary to produce its direct requirements $E_D^{syst}$).

With the notations of Fig. 2.1.1 the correspondence between the two points of view is:

$$\begin{array}{rcl} \textit{syst} \text{ as an auxiliary of } M & & \textit{syst} \text{ as the system under study} \\ E_D^M & \equiv & L_{out}^{syst}, \\ \text{and } E_{EP}^M & \equiv & E(L_{in}^{syst}) + E_{EP}^{syst}. \end{array}$$

Although the quantities $E_{EP}^M$ and $E_{EP}^{syst}$ correspond both to the same flow $E_{EP}$ in the diagram of Fig. 2.1.1, they differ as they are not associated to the same system. At the right end of the equivalence the flow $E_{EP}$ corresponds only to primary energies consumed by the system *syst via* its direct requirements $E_D^{syst}$, whereas at the left end it includes all energy inputs of *syst*, including the heat value $E(L_{in}^{syst})$ of its feedstock $L_{in}^{syst}$ when existing.

Hence the coefficient of conversion $\beta_{syst}$ of the system as auxiliary of system $M$ is:

$$\beta_{syst} = \frac{E_{EP}^M}{E_D^M} \equiv \frac{E(L_{in}^{syst}) + E_{EP}^{syst}}{L_{out}^{syst}}.$$

Or, expressed with the rate $R^{syst}$ defined by eq. 2.1.2, the relation becomes:

$$\beta_{syst} \equiv \frac{E(L_{in}^{syst})}{L_{out}^{syst}} + R^{syst}. \quad (2.1.3)$$

When its output is only a material or a service, the system *syst* is a non energy industry like the phone manufacture, the food industries or the information and communication technologies. Its feedstock input $L_{in}$ is then material (iron ore, agriculture plants...or human time of conceptions), and the quantity $E(L_{in}^{syst})$ is thus zero. Both flows $E_{EP}^M$ and $E_{EP}^{syst}$ coincide. Hence the coincidence between the coefficient of conversion $\beta_{syst}$ for the system used as an auxiliary and its own $R^{syst}$:

$$\beta_{syst} \equiv R^{syst} \quad (E(L_{in}^{syst}) = 0) \quad (2.1.4)$$

Otherwise, when the output of the system represents some form of energy - work, electricity, fuels, steam, or even $NH_3$... -, its inputs are also forms of energy. Either they are included in the feedstock $L_{in}^{syst}$ - as for the agro-ethanol industry -, and $\beta_{syst}$ is defined by eq. 2.1.3, or they are included in the consumptions $E_{EP}^{syst}$ - as in the case of the ammonia industry or the steam engine -, and $\beta_{syst}$ is then defined by eq. 2.1.4.

The expression of $\beta_{syst}$ thus depends on the definition of $R_D^{syst}$ and whether the resource $L_{in}^{syst}$ has a heat value.

The steam engine and the coal mine offer examples of both situations. Moreover they show an interesting problem of interdependence between two industrial systems. Indeed if the steam engine supplies the mechanical energy of coal extraction and cleaning, in turn the coal industry provides the fuel of steam engines.

In the case of the steam engine *st* the practical definition of its requirements $E_D^{st}$ includes all the coal the engine directly uses. The quantities $L_{in}^{st}$ and $E(L_{in}^{st})$ are thus zero. The auxiliary system providing the fuel $E_D^{st}$ of the engine is the coal mine *mine*, which consumes a total amount of coal $E_{EP}^{st}$ to achieve the service ($E_{EP}^{st} = \beta_{mine} \cdot E_D^{st}$). The factor $\beta_{st}$ for the steam engine as an auxiliary is defined by eq. 2.1.4:

$$\beta_{st} \equiv R^{st} = \frac{E_{EP}^{st}}{L_{out}^{st}} = \frac{\beta_{mine} \cdot E_D^{st}}{L_{out}^{st}}.$$

In the case of the coal mine *mine* (including its auxiliaries like steam engines), the quantities $L_{in}^{mine}$ and $E(L_{in}^{mine})$ represent the part of raw coal to be cleaned and sold. The energy input $E(L_{in}^{mine})$ thus coincides with the heat value of $L_{out}^{mine}$. The flow $E_{EP}^{mine}$ represents the remaining part of raw coal consumed on-site to power the engines, which supply the mechanical requirements $E_D^{mine}$ of the mine ($E_{EP}^{mine} = \beta_{st} \cdot E_D^{mine}$). Hence, from the point of view of the coal industry as an auxiliary system, the factor $\beta_{mine}$ is defined by eq. 2.1.3:

$$\beta_{mine} \equiv \frac{E(L_{in}^{mine})}{L_{out}^{mine}} + R^{mine} = \frac{E(L_{in}^{mine}) + E_{EP}^{mine}}{L_{out}^{mine}} = \frac{E(L_{in}^{mine})}{L_{out}^{mine}} + \frac{\beta_{st} \cdot E_D^{mine}}{L_{out}^{mine}}.$$

The right hand expressions of the two above equations show the interdependence of efficiencies between mine and steam engine. For the mine as auxiliary and by using the direct rates, $R_D^{st} = E_D^{st}/L_{out}^{st}$ and $R_D^{mine} = E_D^{mine}/L_{out}^{mine}$, which express the very specific and directly available efficiency of each system (or its main operations), the factor $\beta_{mine}$ is actually:

$$\beta_{mine} = \frac{1}{1 - R_D^{st} \cdot R_D^{mine}}.$$

We have assumed $L_{out}^{mine} = E(L_{in}^{mine})$. When the term $R_D^{st} \cdot R_D^{mine}$ is larger than one, the self-reliant system does not work. The present resolution corresponds not only to an academic exercise, but also to the situation encountered in deep mines operating with the first steam engines (as exposed in the section 5.1.2 of chapter 5).

The following section about energy systems defines a less ambiguous rate, $R_{diss}^{in}$ (eq. 2.2.1), and hence a unique factor $\beta_{syst}$ whatever the energy system.

## 2.2. Case of Energy Systems

The outputs of an energy system, as well as its inputs, are or could be valued for its energy content. Examples of such systems are the steam engine, the electricity production, the agro-ethanol production and all primary energy systems like the coal and the petroleum industry. The ammonia manufacture and the heat pump can also be considered as energy systems.

Until now, to establish the practical rates $R_D$ and $R_{EP}$ we have not distinguished between an energy system and a non energy or material one. We have just mentioned the ambiguity existing between the inputs $E_D$ and $L_{in}$, which introduces also an ambiguity in the definition of the rates $R_D$ and $R_{EP}$ (see section 2.1.1).

Moreover, any energy system and its flows must obey the thermodynamics laws. Thus the system cannot use more energy than that is flowing *via* all its inlets. Can we define an indicator of efficiency or a rate which expresses this important limitation?

To advance towards this aim we first present a new diagram of inputs and outputs better suited to energy systems. From this diagram and the one in Fig. 2.1.1, we examine the different possibilities to define a rate of consumption. The energy conservation is finally applied to provide a definition of an efficiency indicator for these systems.

### 2.2.1. Inputs and Outputs of an Energy System, and Its Possible Rates

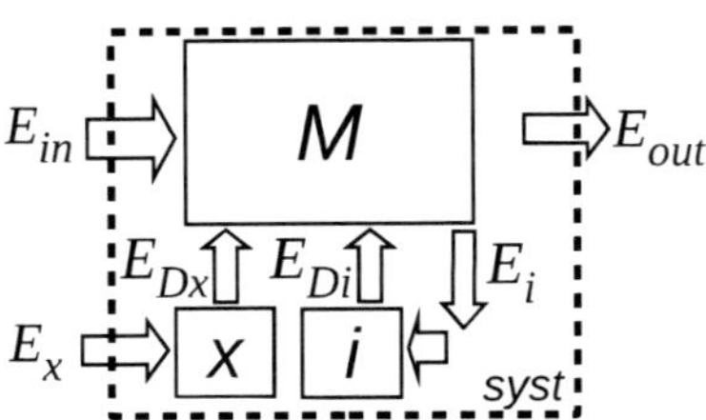


Figure 2.2.1. Typical diagram of flows for an energy system *syst* producing the output $E_{out}$ from its main input $E_{in}$. Direct energies and materials $E_{Dx} + E_{Di}$ consumed by the system main operations $M$ are provided by external energies $E_x$ for the part $E_{Dx}$, and by the energies $E_i$ of the system itself for the part $E_{Di}$, *via* the auxiliary operations $x$ and $i$ respectively.

**Diagram of Flows** The diagram of flows in Fig. 2.2.1 models more precisely than the diagram in Fig. 2.1.1 an actual energy system.

The system net output $E_{out}$ represents the input $L_{out}$ of Fig. 2.1.1 converted into its heat value (a subscript in the unit indicates its form). Thus the steam engine output, 1 $m^3$ over a height of 1 m, is worth about 9 800 $J_m$ of mechanical energy. In the case of the ammonia industry, 1 kg of $NH_3$ presents a HHV of 22.5 $MJ_{HHV}$. One $m^3$ of ethanol at the ambient condition has a low heat value of 21.3 $GJ_{LHV}$.

Likewise the main input $E_{in}$ of the system, which is first used as feedstock to produce $E_{out}$, is expressed in an energy unit.

It does not necessarily coincide with the input $L_{in}$ of Fig. 2.1.1. When considering a coal mine operating with a steam engine, the energy in the input $E_{in}$ of Fig. 2.2.1 aggregates both the coal to be sold after cleaning and the coal burnt by the engine for the mine operations. But in Fig. 2.1.1 the quantity $L_{in}$ represents only the coal to be processed as feedstock, while the mine requirements $E_D$ are provided by the engine from the coal burnt as fuel $E_{EP}$. In the case of the mine, the flow $E_{in}$ of Fig. 2.2.1 thus includes the quantities $L_{in}$ and $E_{EP}$ of Fig. 2.1.1. On the other hand, in the case of the agro-ethanol industry both quantities $E_{in}$ and $L_{in}$ refer to the whole of cereals or sugar plants processed into ethanol.

In the flow diagram of Fig. 2.2.1 a part $E_i$ of the input $E_{in}$, or of its successive products in the system - including a part of the gross output -, is diverted to provide the internal fraction $E_{Di}$ of the direct requirements $E_D$ of the system. An auxiliary system $i$ may be necessary to produce the usable form $E_{Di}$. The other fraction $E_{Dx}$ of $E_D$ is produced from the consumption $E_x$ of external primary energies, of which forms differ from the main input.

The coal powered steam engine and the gas based ammonia industry use their main input both as their feedstock and fuel. External energy $E_x$ is thus zero.

Past exploitation of coal in la Loire basin used a small part of its input to provide internally all its requirements. Flow $E_x$ was also zero.

The corn based agro-ethanol industry satisfies presently all its requirements with external energies. Internal energy $E_i$ is zero. The sugar cane based agro-ethanol industry burns the cane fiber to satisfy part of its requirements. It requires both consumptions $E_i$ and $E_x$.

In the case of the heat pump, the input - some heat of ambient air - represents a too degraded form of energy to be useful for providing the machine requirements $E_D$. All of them are supplied externally from the electricity grid. The diagram of flows in Fig. 2.1.1 is sufficient to describe this system.

It should be noticed that the status of a direct requirement, that is either from internal sources - $E_{Di}$ - or external sources - $E_{Dx}$ -, depends at which level it is considered, the system or the operation where it is consumed. The transport by oil tanker within the petroleum industry requires the heavy fuel produced by the refineries of the system. The fuel is thus considered an internal energy. On the other hand, when the transport is studied separately the fuel becomes an external energy.

**Determination of a Rate *R* for Energy Systems** A first choice consists in applying eq. 2.1.2 based on the flows in Fig. 2.1.1. By identifying these flows with that in Fig. 2.2.1, two solutions are possible:

$$E_D \equiv E_{Dx} \quad \text{and} \quad E_{EP} \equiv E_x$$
$$\text{or } E_D \equiv E_{Dx} + E_{Di} \quad \text{and} \quad E_{EP} \equiv E_x + E_{in} - E_{out}$$

In the latter case the flow $E_{EP}$ includes the external and internal consumptions. The difference $E_{in} - E_{out}$ is preferred to the quantity $E_i$ to take into account all the internal consumptions, i.e. the fraction of the input $E_{in}$ internally dissipated.

In terms of rates the two possibilities give:

$$R_x^{out} = \frac{E_x}{E_{out}} \quad \text{or} \quad R_{x+i}^{out} = \frac{E_x + E_{in} - E_{out}}{E_{out}}$$

The difficulty to define a rate is also compounded by the choice of the rate denominator. Instead of the output $E_{out}$, we can chose the input $E_{in}$:

$$R_x^{in} = \frac{E_x}{E_{in}} \quad \text{or} \quad R_{x+i}^{in} = \frac{E_x + E_{in} - E_{out}}{E_{in}}$$

Thus for the same system and its flows the value of the rate is dependent on its definition.

**Example of the Corn Based Ethanol Industry in the USA** The two diagrams of Fig. 2.2.2 show graphically, in the case of ethanol production from corn, the influence of the definition of the rate independently of the validity of flow data. In each diagram the energy flows of the system are expressed like in Fig. 2.2.1 but in such a way as to fix at 100 J the value of the flow used to define the rate denominator - either $E_{out}$ in Fig. 2.2.2(a) or $E_{in}$ in Fig. 2.2.2(b) -. The choice of the consumption flows in the numerator imposes then the definition of the rate, $R_x^{out}$, $R_{x+i}^{out}$, $R_x^{in}$ or $R_{x+i}^{in}$.

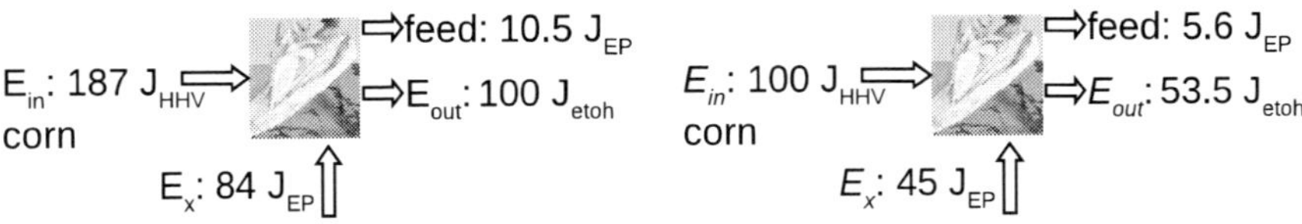


(a) Energy flows for the production of 100 $J_{LHV}$ of ethanol (rates $R_x^{etoh}$ and $R_{x+i}^{etoh}$).

(b) Same data as in Fig. (a) but for an input of 100 $J_{HHV}$ of corn (rates $R_x^{in}$ and $R_{x+i}^{in}$).

Figure 2.2.2. Main - corn - and external energy inputs - primary energies -, and outputs - ethanol and feed - for the production of ethanol in the USA in 2007.

The first rate $R_x^{out}$, or more precisely $R_x^{etoh}$, takes into account only the net consumption of external primary energy (the gain from feed production has been subtracted). It is the rate of choice in the literature to study the efficiency of the agro-ethanol industry. It corresponds also to the practical rate defined by eq. 2.1.2 with $E_{EP} \equiv E_x$ and $L_{out} \equiv E_{out}$.

The rate amounts to about 74 $J_{EP}$ for 100 $J_{LHV}$ of ethanol produced after subtraction of feed gain. For information let's give some details or breakdown of this quantity, according to our study of the system [50] based on the methodology exposed in the chapter 4. For 100 $J_{LHV}$ of ethanol a consumption of 17 $J_{EP}$ is required at the farm, around 3 $J_{EP}$ for the transport of corn by trucks, 62 $J_{EP}$ in an average factory - operating with no cogeneration, and concentrating and drying the stillage to produce an animal feed -, and 1.5 $J_{EP}$ for the ethanol distribution to the dispenser. These values correspond to averages over typical processes in 2007. Costs due to irrigation, like in Nebraska, are excluded. Modifications and improvements of the processes make other values of $R_j^{etoh}$ possible, especially at the factory [50,51]. The total expenditure $E_x$ of external energies along the industry chain represents 84 $J_{EP}$ for 100 $J_{LHV}$ of ethanol. As a by-product - a protein rich animal feed - is made along with ethanol, we must attribute it an energy value. Because it is not burnt to produce a heat like fuels, instead of its heat value we consider the primary energy saved by not producing the feed for which it is substituted.

The second rate defined with respect to ethanol, $R_{x+i}^{etoh}$, includes, besides $E_x$, the internal dissipation, $E_{in} - E_{etoh}$. The resource input $E_{in}$, 187 $J_{HHV}$, is deduced from the high heat value HHV of the corn, knowing its constituents - mostly starch - and its moisture - 15% in average. The rate $R_{x+i}^{etoh}$ amounts to 160 $J_{EP}$ for 100 $J_{LHV}$ of ethanol produced.

The same data can be presented with the plant input instead of the ethanol output for the rate denominator, as shown in Fig. 2.2.2(b). Hence two rates ares possibles:

- The net consumption $R_x^{in}$ of external primary energy to process 100 $J_{HHV}$ of corn into ethanol, 39.5 $J_{EP}$.
- The consumption $R_{x+i}^{in}$ including both external and internal energy use to process 100 $J_{HHV}$ of corn, 86 $J_{EP}$.

Among these four rates some present such a deficiency that their use can be misleading. Let's imagine a system for which $E_{in}$ provides both for the feedstock and the energies consumed directly or indirectly (such as the sugar cane in the agro-ethanol factory). With the definitions in form of $R_x$, excluding internal use, the rates amount to zero and nothing can be said about the system efficiency.

Consequently, the rates $R^{out}_{x+i}$ or $R^{in}_{x+i}$ appear as better indicators. However, these rates are not consistent with the energy conservation. Their value can be higher than 100 J, as observed for $R^{out}_{x+i}$ in the ethanol case.

### 2.2.2. Energy Conservation

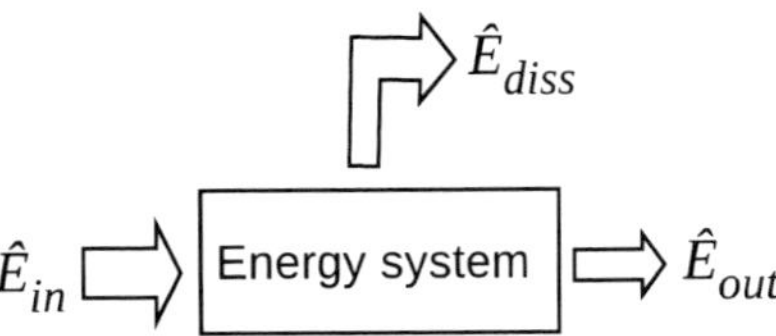


Figure 2.2.3. Schematic flow diagram in a closed energy system. $\hat{E}_{in}$ is the energy input of the system, while $\hat{E}_{out}$ is its output and $\hat{E}_{diss}$ the energy dissipated by its operations.

**Closed systems** Fig. 2.2.3 represents schematically a closed system *syst* that processes an amount $\hat{E}_{in}$ of energy over a fixed period. A fraction $\hat{E}_{diss}$ is dissipated or consumed in the operations to produce the useful part $\hat{E}_{out}$. The part $\hat{E}_{diss}$ is not anymore usable (the waste heat when recovered are thus excluded). The efficiency of this simple system can only be defined by the following rates, which are absolute indicators of dissipation:

$$R^{out}_{diss} = \frac{\hat{E}_{diss}}{\hat{E}_{out}} \text{ or } R^{in}_{diss} = \frac{\hat{E}_{diss}}{\hat{E}_{in}}.$$

By energy conservation we have $\hat{E}_{in} = \hat{E}_{diss} + \hat{E}_{out}$. As a result the above rates can be expressed by:

$$R^{out}_{diss} = \frac{\hat{E}_{in}}{\hat{E}_{out}} - 1 \text{ or } R^{in}_{diss} = 1 - \frac{\hat{E}_{out}}{\hat{E}_{in}}.$$

The latter form is strictly higher than 0 and lower than 1; the closer the rate is to 1, the more inefficient is the system. The former form varies strictly between 0 ($\hat{E}_{diss} = 0$ or $\hat{E}_{out} = \hat{E}_{in}$) and $\infty$ ($\hat{E}_{diss} = \hat{E}_{in}$ or $\hat{E}_{out} = 0$); the larger the rate, the more inefficient is the system.

From the previous relations the rate $R^{out}_{diss}$ can be expressed with respect to $R^{in}_{diss}$:

$$R^{out}_{diss} = \frac{R^{in}_{diss}}{1 - R^{in}_{diss}} .$$

The form $R^{in}_{diss}$ is preferred thereafter in order to provide a better control of potential errors ($R^{in}_{diss}$ must be lower than 1).

In summary the rate of dissipation of a closed system is:

$$R^{in}_{diss} = \frac{\hat{E}_{diss}}{\hat{E}_{in}} = 1 - \frac{\hat{E}_{out}}{\hat{E}_{in}} . \tag{2.2.1}$$

From these relations, other quantities are deduced. When used as an auxiliary, the system provides for the direct requirements with a conversion $\beta_{syst}$:

$$\beta_{syst} = \frac{\hat{E}_{in}}{\hat{E}_{out}} = \frac{1}{1 - R^{in}_{diss}} . \tag{2.2.2}$$

When working with the yield $Y$ rather than the rate to define the system efficiency:

$$Y = \frac{\hat{E}_{out}}{\hat{E}_{in}} = 1 - R^{in}_{diss} = \frac{1}{\beta_{aux}} . \tag{2.2.3}$$

**Correspondence to an Actual Energy System** A general energy system *syst* described by the diagram of Fig. 2.2.1 can be assimilated to a closed system with:

- $\hat{E}_{in} \equiv E_{in} + E_x$,
- $\hat{E}_{diss} \equiv E_{in} - E_{out} + E_x$,
- $\hat{E}_{out} \equiv E_{out}$.

In terms of rate, we would have the following equivalence:

$$R^{in}_{diss} \equiv R^{in+x}_{i+x}$$

Let's establish the rate $R^{in}_{diss}$ in the case of the gas fueled ammonia $NH_3$ industry.

Its practical rate $R_D^{NH_3}$ is defined as the direct gas consumption to produce one tonne of ammonia. The gas represents all the input,i.e. $\hat{E}_{in}$(Fig. 2.2.3)$\equiv E_{in}$(Fig. 2.2.1)$\equiv E_D$(Fig. 2.1.1) - gas processing and transport are not taken into account -. The output $\hat{E}_{out}$(Fig. 2.2.3)$\equiv E_{out}$(Fig. 2.2.1)$\equiv E(L_{out})$(Fig. 2.1.1) corresponds to $NH_3$, of which heat value is 22.5 $GJ_{HHV}.t_{NH_3}^{-1}$ [41].

Hence the part of the gas per tonne of ammonia used to fuel the processes, and so dissipated, is $R_{diss}^{NH_3} = R_D^{NH_3} - 22.5\ GJ_{HHV}.t_{NH_3}^{-1}$.

The rate $R_{diss}^{in}$ of dissipation defined by eq. 2.2.1 represents the ratio of $R_{diss}^{NH_3}$ to $R_D^{NH_3}$.

The rate $R_D^{NH_3}$ for the American industry amounts to 38 $GJ_{HHV}.t_{NH_3}^{-1}$ [21]. The quantity dissipated is thus $R_{diss}^{NH_3} = 15.5\ GJ_{HHV}.t_{NH_3}^{-1}$. Hence the rate $R_{diss}^{in}$ for the American industry is $R_{diss}^{in} = 41\%$.

As far as the best factories are concerned the respective rates are $R_D^{NH_3} = 30\ GJ_{HHV}.t_{NH_3}^{-1}$ and $R_{diss}^{NH_3} = 7.5\ GJ_{HHV}.t_{NH_3}^{-1}$. Hence the rate $R_{diss}^{in} = 27\%$.

In this example the rate $R_{diss}^{in}$ enhances more than the practical rate $R_D^{NH_3}$ the gains of efficiency.

If the energy dissipated in upstream operations of the gas industry must be taken into account, it would be included in both the numerator and denominator of the rate $R_{diss}^{in}$. Thus it increases the value of $R_{diss}^{in}$ as expected.

We can also determine the rate $R_{diss}^{in}$ of dissipation for the agro-ethanol industry in the USA studied in the section 2.2.1 (Fig. 2.2.2). Its flows are presented such as to fix at 100 $J_{HHV}$ the sum of the inputs of corn and external primary energy, $\hat{E}_{in} \equiv E_{in} + E_x$, (Fig. 2.2.4). Hence, according to eq. 2.2.1 and the correspondence of flows between Fig. 2.2.3 and 2.2.1, the rate $R_{diss}^{in}$ amounts to 59 $J_{EP}$.

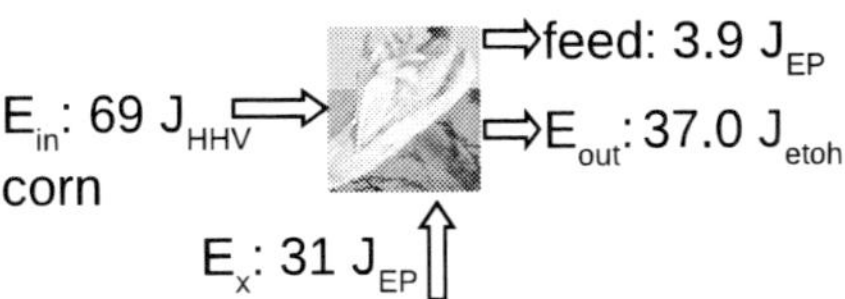


Figure 2.2.4. Same as Fig. 2.2.2 with the sum of the inputs of corn and external energy fixed at 100 $J_{HHV}$ (case of rates $R_{i+x}^{in+x}$ and $R_{diss}^{in}$).

**Self-Reliant Systems** A particular case of an energy system is the self-reliant system corresponding to $E_x = 0$ in Fig. 2.2.1. The system consumes part of its feedstock, directly or indirectly (successive forms in the chain of operation), to fuel its operations.

Simple systems like ammonia industry or steam engine consume the same material as their feedstock and their fuel, and consequently can be considered self-reliant. More complex systems are illustrated in the chapter 5 with an ancient coal exploitation, self-reliant by necessity. The petroleum industry is probably among current energy systems the one closest to self-sufficiency [20].

The section 5.2.1 of the chapter 5 shows in the case of the agro-ethanol industry in the USA how to convert a present industry into a self-reliant one. Actual processes substitute for external processes $x$ of the auxiliaries in order to use internal forms of energy. Hence the rate $R^{in}_{diss}$ of the new system would not include the efficiency of external systems and their feedstock.

However, the new auxiliary processes may require such a quantity of energy that all the input of the system is consumed to provide it. The system does not produce any output (like the very deep coal mines operating with the first steam engines in the chapter 5). Consequently it cannot exist without the feedstock and efficiency of other external systems.

## 2.3. Other Criteria to be Considered for an Industrial System

### 2.3.1. Constraints on Process Choice

An important point of the study is that the system and its processes, and consequently the related data to derive the rate $R$, correspond to an actual situation tested at least in an industrial pilot. Virtual heat engines and their yield $Y = 1 - R^{in}_{diss}$ obtained from the thermodynamic cycles according to Carnot's work are rarely representative of a real process. It may be an indication of further progress, but other limits often impede this progress (constraints from available working fluids and/or hot sources, material resistance at high temperature and pressure, and in the corrosive environment of the fluids, safety risks...).

Thus the Stirling machine - an enclosed gas in one phase cycling between two external sources and a heat regenerator - may present a theoretical yield equal to Carnot's one - yield $Y$ of 78% between a hot source at 1000°C and a cold one at 0°C, or a rate $R^{in}_{diss}$ of 22% -. But in industrial applications this engine $Y$ is limited to 20% due to the difficulties for a correct heat transfer between the sources, the regenerator and the gas, to the leaks, to its limited flexibility...

The same critic can be formulated to the exergy notion. The exergy corresponds to the work extracted from the heat through an ideal Carnot cycle of which the cold sink is the outside environment at a temperature around 15°C [5]. It presents an interest to estimate the useful value of a form of energy, like heat according to its temperature, and to track the irreversibilities of a chain of processes.

The comparison between theory and application is studied in details in the case of the steam engine (section 5.1.1 of the chapter 5).

Nevertheless, the theoretical framework is useful for the variables and fundamental limit it provides to the rate of dissipation $R^{in}_{diss}$.

Interestingly, S. Carnot at the end of his book warned against a too much focus on the search of the maximum of efficiency. It can be detrimental to other necessities such as safety (the explosion of some steam engines occurred with the development of steam at higher pressures), durability, and low space requirements and costs. Thus the system under study must meet some constraints, which may impact its energy consumptions:

- productivity in terms of labor and expenditures per unit of output,
- quality of products (reliability, durability, or purity...) for its commercial interest,
- safety of the equipments of production,
- neutralization of the contaminants produced during the process ($NO_x$, $SO_x$, organic residues, particles...).

Large scale production usually decreases the economic cost per unit of product, especially by reducing the total number of labor hours and/or by a better utilization and, hence, amortization of the equipment. The scale factor is less pronounced in energy efficiency. The energy cost to make a photovoltaic panel as well as its electricity output are both proportional to the panel surface. Hence the rate $R$ of the system is independent of the surface. However, all systems in which energy consumptions are proportional to the surface of its equipment and the production to the volume such as for a furnace or for the transport in a tanker, can present a rate $R$ decreasing with the size. The larger the Newcomen steam engine was, the more efficient it was. In 1765 J. Watt put in evidence its intrinsic inefficiencies thanks to a small sized model of the engine. The length of equipments thus becomes a physical variable, which the methodology suggested in chapter 4 must show.

Low capital expenditure can impact the energy consumption with the choice of equipments of poorer performance than more expensive ones. For this reason the Newcomen machines were still manufactured at the beginning of the nineteenth century in spite of the existence of more efficient Watt engines [1,2]. They were also found easier to maintain.

Controls of the product quality, the safety of the processes and the pollution as well as the labor productivity can require additional operations and/or more sophisticated processes with higher energy consumptions. The rapid and round the clock availability of a service necessitates a permanent consumption (equipments in standby or idle modes, scheduled routes of buses even at off-peak hours). As all consumptions are reported against the service actually performed, the efficiency is all the smaller that the service is poorly used (the factor of utilization is thus a variable of the rate $R$ of the service).

As far as the contaminants are concerned, the contribution of the energy consumptions to neutralize them is all the lower that they represent small quantities relative to the output of the system main product. Besides the quantities are very sensitive to the choice of processes. Hence the rates of emission of the contaminants have uncertainties larger than that for the energy consumption rates.

### 2.3.2. Indicators of the Criteria in Form of Rates of Use or Occurrence

The economic, safety and environmental criteria described above can be expressed in quantities per unit of output like in the case of energy consumptions with eq. 2.1.1 and 2.1.2:

- materials in investment and operation per unit of $L_{out}$,
- worker.hours per unit of $L_{out}$,
- \$, ¥, or € in operational and capital expenditures per unit of $L_{out}$.
- number of breakdowns of products over a fixed time per unit of $L_{out}$,

- number of accidents and/or deaths per unit of $L_{out}$,
- mass of contaminants per unit of $L_{out}$,
- amount of greenhouse gases emitted per unit of $L_{out}$.

Consequently these rates can be obtained thanks to the same methodology exposed in the book for the rate of energy consumption.

Moreover, the direct flow $E_D$ in eq. 2.1.1 is also used to provide the list of materials required by the system. They can be then converted into their monetary values. Furthermore, the list can be useful to estimate the impact of the industry under study in the exploitation of elements from the earth crust.

The rate of the total number of hours worked in worker.hours and the rate of the quantity of purchased materials correspond to physical productivities. They differ from the productivities studied in the section 3.1 of the chapter 3, as the latter ones are expressed in monetary values (price or added value).

Like for the energy consumption, the methodology must determine the flows of the quantity under study both in the main system and in the external systems which provide the requirements of the main system. After determining the system own contribution, the work consists in the collection of the goods or services it requires from other systems. Hence the contributions of the latter to the quantity under study are derived.

**Energy Consumptions and $CO_2$ Emissions** If indeed a close cause/effect relationship exists between the observed recent global warming and the also undoubted increase of concentration of $CO_2$ and other greenhouse effect gases in the atmosphere - due to the use of fossil fuels by humans -, the balance of emissions of these gases from the system under study may be necessary. As for the energy consumed the rate of emission has a better meaning than the absolute flow, when it comes to compare the emissions of various systems with the same type of output.

However these emissions are not measured directly and are mostly derived from an energy balance.

For instance the quantity of $CO_2$ emitted per kWh of electricity from a coal fired power plant is deduced from the electrical yield $Y_e$ and the heating value of the pure coal used. The average yield of the coal fueled power plants in the USA for 2007 was $Y_e = 0.342\ J_e.J_{HHV}^{-1}$ [16]. The H and C atoms of a dry coal present a high heat value of 36 $MJ_{HHV}.kg^{-1}$ with 0.5 to 0.9 atom of H for one atom of C. Consequently the combustion of 1 kg of H and C atoms from the coal will produce simultaneously 3.4 kg of $CO_2$ and 36 $MJ_{HHV}$ of heat, or 12 $MJ_e$ of electricity by taking into account the yield $Y_e$. Hence the rate of emission of $CO_2$ is about 1.0 $kg_{CO_2}/kWh_e$ for the power plants in the USA.

It should be noted that the $CO_2$ emission depends even more on the performances of the external systems than the quantity $E_{EP}$ does. For instance an aluminum can of 15 g requires a quantity of electricity of about 0.9 $MJ_e$ if made from bauxite, or 0.06 $MJ_e$ if it is produced from collected scraps. To derive a carbon balance we must also take into account the system of production of electricity. In Norway 98% of electricity is produced from waterfalls (resulting in about 9 $g_{CO_2}.MJ_e^{-1}$), whereas in Poland electricity is produced from coal at 95% and from heavy oil for the remaining (resulting in about 280 $g_{CO_2}.MJ_e^{-1}$). Depending on the process to manufacture aluminum and on the electricity system of the country where the factory is located, the rate of emission for the electricity consumed to produce the can varies from 0.5 to 240 $g_{CO_2}$.

As the most strenuous part of the work is to obtain the complete energy balance of an industrial system, the present study is only dedicated to methods to achieve this balance. The study should be enough detailed, type of energies..., to derive straightforwardly the carbon balance with the energy structure of the country where the operations take place.

## 2.4. Summary of the Chapter

Whatever the industrial system, its main output and its direct consumptions (Fig. 2.1.1), a practical indicator of its efficiency based on the available raw data of the system flows is defined as the consumptions per unit of output, the rate $R_D$ of eq. 2.1.1. It is converted into a consumption rate of primary energy, the rate $R$ of eq. 2.1.2, thanks to the efficiency of the auxiliary systems providing the direct consumptions from primary energy.

Because part of the resource used as feedstock by the system can fuel the processes and/or is dissipated internally (Fig. 2.2.1), the previous definitions can omit large contributions to the consumptions, especially for the energy systems which produce a form of energy. Thanks to the energy conservation, a single indicator can be defined, the rate of dissipation according to eq. 2.2.1.

However, an indicator of efficiency truly specific to the system under study should correspond to the rate of a self-reliant system. Processes exist to make a current system self-reliant, but with reduced output or not at all.

Industrial system or processes are submitted to other criteria like labor or investment productivities, or contamination control. Their indicators can be expressed in form of physical rates which are obtained using the same methodology suggested to determine the energy efficiency.

# Chapter 3

# Energy Efficiency in the Economy, Population and Resources Extraction

Energy itself is ubiquitous in both the economy and the human lives, although its importance has become less apparent due to the very reliability (at least in developed countries) of the industries providing it. It is however not raw energy that is important to our societies but useful forms of energy (electricity at the grid outlet, refined fuels...). Hence the processes to produce these useful forms and the efficiency of these processes are of high significance. Their fundamental limitations and technical constraints can thus have deep consequences.

Other factors than the energy efficiency are also at play: monetary value of energy, production costs, consumer behavior and resource abundance. This chapter studies their interplay with energy efficiency.

The quantities of energy in this chapter are often reported in the unit of tonne oil equivalent (toe), which value in joule is fixed at 41.9 GJ. Although the Comité International des Poids et Mesures does not recognize this energy unit, it is commonly used in economic statistics and is a convenient way to express and compare energy production and consumption at the level of a country or a person over a year (one toe of energy is a good order of consumption per person and year). Another energy unit used in the chapter is watt.hour, or Wh (usually to express electricity flows but not limited to this form of energy). It represents the energy output of a generator of 1 watt power for one hour, and its value is

3.6 kJ.

## 3.1. Energy Efficiency in the Economy

The mechanisms by which the energy efficiency of a system $M$, as defined in the previous chapter, may impact costs and revenue of the system can be summarized by the following chain of energy, material and monetary flows over a period like a year:

$$\begin{array}{lccccccc} \text{Flows :} & E_{EP} & \underset{\beta_{aux}}{\rightarrow} & E_D\ (\rightarrow \$_D) & \underset{R_D^M}{\rightarrow} & L_{out} & \underset{\text{unit price}}{\longrightarrow} & \$_{\text{out}} \end{array}$$

The consumptions $E_D$ of system $M$ represent the final energies such as petroleum products, processed gas and electricity, as well as materials and chemicals products, which all require some primary energy for their manufacturing in the operations *aux*.

Indicators of efficiency have been defined by the rate of consumption $R_D^M$ and energy conversion $\beta_{aux}$:

$$R_D^M = \frac{E_D}{L_{out}}, \text{ and } \beta_{aux} = \frac{E_D}{E_{EP}}.$$

The flow $\$_{out}$ represents the volume of sales associated to the physical output $L_{out}$ of the industry. It should be put in balance with the industry purchases $\$_D$ associated to energy and material consumptions.

A decrease of the quantities $\beta_{aux}$ and/or $R_D^M$ leads in the long run to a reduction of the costs $\$_D$, and subsequently of the price of products (or should do so). Because natural resources just before their extraction are free and because the financial balance of an economic unit is established from its direct consumptions, the rate $R_D^M$ is mostly used.

The ammonia $NH_3$ industry offers an example of the large progress made during the twentieth century on the rate $R_D^{NH_3}$. It was achieved through application of thermodynamic and kinetic theories, and the use of engineering sciences.

However, these same important fundamental and technical studies do not forecast major further gains for the Haber-Bosh process after one hundred years of active research and development.

Because the unit price of a product ($u\$_{out} = \$_{out}/L_{out}$ or $u\$_D = \$_D/E_D$) can present its own evolution or is actually dominated by non energy related items such as wages, taxes or investment return, economic efficiency of the system defined by monetary values may disconnect from the rate $R_D$. Hence equipment can represent meaningful items in financial terms, even if it is not the case in energy terms (The section 4.3.4 of the chapter 4 provides the example of the tractor at the farm to produce crops for agro-ethanol production).

The recent trends in monetary and energy flows in various industries and sectors of the American economy are examined to find examples of this disconnect.

Another and more thoroughly studied example is provided by the development of the global industry of photovoltaic modules and its market since 2000. At the difference of the previous sectors the industry has presented such a huge growth that its present capacity is at least one thousand time as high as it was at the beginning of the millennium. It has stemmed from both government regulations and huge investment flows. The growth allowed the rapid implementation of last improvements in processes, however, to the profit of a mature technology. As a result of all these developments relationship between revenues, costs and energy efficiency is more intricate. Equations are established between sector profit and all these factors to measure their respective influence.

In the end we deliberate whether the rate $R_D$ can offer a long-term indicator of the future of the economy and its growth.

### 3.1.1. Historic Evolution of the Efficiency of some Primary Industries

#### Case of the Ammonia Industry Based on the Haber-Bosh Process Since Its Inception

**Historic Context and Importance** With the advent of modern chemistry in the 19th century (discovery of elements, atomic theory, organic chemistry...) and its application in industry and agriculture, the crucial role of nitrogen N for plant growth and in subsequently agricultural yields was finally acknowledged [6]. The importance for the assimilation of N by plants of mineral forms such as nitrate and ammonia $NH_3$ was also recognized.

After these discoveries the exploitation of natural nitrate deposits and the search for industrial processes of its production intensified. However, early processes (electric arc in air or reaction of calcium carbide with $N_2$ at high temperature) were expensive and their output limited largely due to their energy inefficiency (Fig. 3.1.1). In 1910 most of the industrial production came from nitrate extracted in Chilean deserts (360 $kt_N$) and from ammonia recovered in the coke ovens (230 $kt_N$). Others sources represented less than 30 $kt_N$ a year.

At the beginning of the twentieth century the German chemist F. Haber clarified the laws which governs the synthesis of ammonia from dihydrogen $H_2$ and nitrogen of air $N_2$ in a gaseous phase:

$$3\,H_2 + N_2 \rightarrow 2\,NH_3$$

He obtained the correct parameters for his theory and, importantly, he achieved a laboratory experiment which could be scaled to an industrial size. From this research C. Bosh and his colleagues at the BASF industry succeeded in 1913 to build the first industrial pilot of the so-called Haber-Bosh process with an output of 8 $t_{NH_3}$ per day. As soon as 1917 the first large factory was able to produce about 35 $kt_N \cdot y^{-1}$.

Since then the production of ammonia has increased constantly to reach about 100 $Mt_N \cdot y^{-1}$ in 2010. Its highest growth rate was attained between 1950 and 1980 with a production rising from 4.7 to 61 $Mt_N \cdot y^{-1}$.

**Overview of the Chain of Processes: Thermodynamics, Kinetics and Other Constraints** A thermodynamic analysis, based on the free enthalpy or Gibbs function of the synthesis reaction, shows that the reaction must take place at high pressure (reduction of the number of gaseous moles) and low temperature (exothermic reaction) [7]. However, at a typical pressure of 150 bars and temperature of 450°C, the theoretical yield of conversion is only 50%. Moreover, due to kinetic constraints it would require a prohibitive time delay. The lower the temperature is, the higher the theoretical yield but the longer it takes to reach it.

A catalyst was essential for a meaningful output, even at laboratory scale. Chemists of BASF screened more than two thousand minerals to select the most active catalyst at a reasonable cost [8]. But, even with the best available catalyst, the reaction has to operate at sufficiently high temperature to activate it. With the above thermodynamic conditions, the selected catalyst and a residence time of one minute, the practical yield is close to 20%. To recover unused reagents F. Haber and C. Bosh had to conceive a recirculation loop which includes a separation of $NH_3$.

They also faced with the difficulties to build a large reactor resilient enough to withstand pressures higher than 100 or 200 bars as well as the chemical action of $H_2$ (ordinary carbon steel becomes brittle). Moreover heat exchangers were required to control the temperature.

However, in 1915 the engineers succeeded in operating the first reactor with an output of 85 $t_{NH_3}$ per day [6].

In the ammonia factory the nitrogen $N_2$ is provided by air, whereas $H_2$ must be produced from natural feedstock, coal in early time, later the oil residues and today from natural gas. Consequently the factory includes the reactions to produce $H_2$ from the feedstock, partial oxidation and/or steam reforming. They produce other molecules (S compounds, CO, $CO_2$, $H_2O$, $CH_4$...), some of which are poisonous for catalysts (S compounds, CO, $CO_2$ and $H_2O$ should be only present in very tiny traces). To remove them and to adjust the gas composition, a long chain of processes precedes the actual synthesis: desulfurization, water shift process to consume most of CO, $CO_2$ removal, methanation to eliminate traces of CO and $CO_2$. Each of them has its own thermodynamic and kinetics constraints, which imply limited ranges of temperature and pressure, and the use of catalysts for some reactions. As a result changes of pressure and temperature between each reaction, and a judicious order of them to avoid catalyst poisonous, are often required.

**Gains and Limitations on Energy Efficiency and Other Productivities** The Fig. 3.1.1 shows through the changes of processes and their improvements the evolution of the rate $R_D^{NH_3}$ of the overall energy required to produce $NH_3$ in an ammonia factory. The best factories operating with natural gas currently consume only 34 $MJ_{LHV}\cdot kg^{-1}_{N(NH_3)}$, or 30.5 $MJ_{HHV}\cdot kg^{-1}_{NH_3}$ (using the conversions between gas HHV and LHV, and between N and $NH_3$ molar masses) [8, 10].

Large gains were obtained through process integration with the recovery of waste heat from exothermic reactions (like the ammonia synthesis itself) to be consumed in endothermic reactions, to preheat the air and fuels, or to heat water and superheat steams [8]. Since Haber and Bosh first reactors, various sophisticated heat exchangers have been conceived and built for the synthesis reaction itself.

The use of natural gas, which is cleaner and has a higher H content than the coal or heavy fuel oils, also reduced the importance of upstream reactions and so their consumptions. It has allowed the substitution of some of these reactions with more efficient ones.

At the same time larger synthesis reactors can be manufactured affording economies of scale. In 1955 the nominal output of a single reactor had slowly

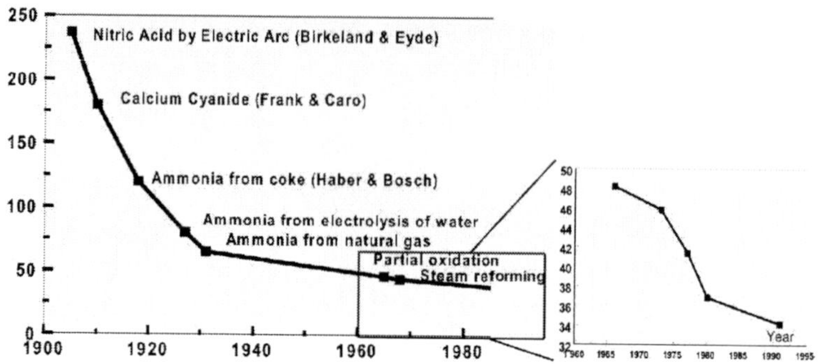


Figure 3.1.1. Evolution during the twentieth century of the rate $R_D^{NH_3}$ in $MJ_{LHV}^{-1} \cdot kg_{N(NH_3)}^{-1}$ for the $NH_3$ industry with the advent of new processes, their improvements and the use of better feedstock (source: G. Kongshaug [9]).

increased to 220 $t_{NH_3}$ per day. Owing to progresses in material quality it reached a capacity of 1 200 $t_{NH_3} \cdot d^{-1}$ in 1970. Typical reactors today have a capacity of 2 200 $t_{NH_3} \cdot d^{-1}$.

Gains of productivities in labor and capital investments have been achieved together with that of energy efficiency. The steel requirements to build the production facility per unit of capacity decreased from 38 t in 1940 to 7 t per $t_{NH_3} \cdot d^{-1}$ in 1991. Labor productivity rose even more dramatically during this period, from 2.25 to 0.055 workers per $t_{NH_3} \cdot d^{-1}$.

However, opposing constraints resulted in lower and lower gains over time.

For a good recovery of a heat potential, exchange between two flows have to take place with a small difference of their temperatures, according to Carnot principle of maximal yield of reversible transformations (see Section 5.1.1 of the chapter 5). However, this difference is constrained by operating temperatures of streams and reactions towards which heat is transferred. Reactions operating temperatures are themselves determined by a compromise between thermodynamics, kinetics and resistance of materials. Thus the steam reforming of gas to produce $H_2$ is an endothermic reaction operating at high temperature (more than 600°C even with a catalyst). Even if its inputs are preheated, it requires high temperature heat produced by an exothermic reaction such as the partial oxidation, which rises energy consumption.

Extra temperature of recovered heat can be lowered while producing works through turbines. However, once achieved, no more gain is available, except through improvements in turbine efficiency.

Moreover, all these gains necessitate sophisticated and expensive equipment. Thus heat transfer at low temperature difference is achieved owing to exchangers with separators presenting large thermal conductivity and surfaces of contact.

Pressure regulation between reactions is subject to similar difficulties. The pressure in the steam reforming reactor must be lower than 50 bars, whereas gas at the factory inlet is provided at 60 bars, and synthesis reaction requires pressure higher than 100 bars. Some expanders save energy by recovering extra pressure work from gaseous streams to produce mechanical work or electricity.

Reduction of pressure and temperature in a reactor allows some improvements in consumption rates. On the other hand, it implies a reduction of the throughput or an increase of the volume of catalyst to keep the same output. Due to larger surface areas, finer grains of catalyst allow a better activation and output. However, they increase the friction of the flow and consequently the energy consumption of the compressors.

Tightening regulations about emissions impose decontamination processes or to lower the process yields. Thus, because $NO_x$ emission results from reaction of air nitrogen at high temperature, its control implies a reduction of the temperature of highly exothermic reactions like gas partial oxidation, and so their usefulness. Alternatively, additional processes like scrubbing eliminate $NO_x$, but at the cost of additional consumptions. Likewise residual $NH_3$ must be removed from effluents.

**What Future Gains are Possible?** The previous examples show that a very large range of savings and recoveries have already been put into practice, taking into account the fundamental limitations and the economic constraints. Within this range of possibilities different designs for a modern gas fueled factory have resulted in close values of efficiency, approaching 30 $MJ_{HHV}{\cdot}kg_{NH_3}^{-1}$ [8, 10].

A rough thermodynamic analysis, based on the free enthalpy of the overall reaction of conversion of methane and air into $NH_3$ - within the Ellingam approximations - and under kinetic constraints, stipulates a theoretical limit of 23 $MJ_{HHV}{\cdot}kg_{NH_3}^{-1}$ [11] (extreme case of the free enthalpy at zero).

Nevertheless, the rate must at least contain the high heat value of one kg of ammonia, 22.5 $MJ_{HHV}{\cdot}kg_{NH_3}^{-1}$.

Accounting for all other sources of consumption/dissipation like heating, compression, pumping to move feedstock and products, $NH_3$ separation in the recirculation loop, neutralization of effluents, packaging... even with energy recoveries, the best factories are today close to the known fundamental limits.

Thus further progress on the rate $R_D^{NH_3}$ of the Haber-Bosh process is expected to remain small relative to present performance.

Is a new catalyst available to reduce the pressure? A mixture with costly ruthenium has increased the $NH_3$ yield. However it represents a small progress in one hundred years of investigation.

The water electrolyze as an alternative to produce $H_2$ is more energy consuming, even with hydroelectric plants operated in Norway. At least this production method offers access to more abundant feedstock than natural gas.

Another possible avenue might be a direct synthesis from water and $N_2$ of air with the use of an electricity powered proton exchange ceramics? The electricity consumption would be lower than with water electrolyze. However no industrial scale pilot has yet been set up.

The limitation of the rate $R_D^{NH_3}$ has significant consequences for the financial profitability of the industry: this one can only be improved by a cheap feedstock or a price rise of its sale. But the latter would be detrimental to the whole economy (see section 3.1.2).

The final improvement for the whole economy will result from the replacement of old factories through modern ones, as in the USA (see below). This, however, is a slow process, especially when the output pricing provides a low return on investments.

### Other Examples of the Influence of Consumption Rates

**From Gas to Proteins *via* Ammonia Production** Let's examine the development of the efficiency of the entire chain of operations between the natural gas in its reservoir rocks and the end products, human food protein produced by modern industrial agriculture.

From the gas to the vegetable proteins we can look at the chain of operations relevant to the area of industrial processes. Gas extraction, processing and transport relate to mature technologies like the ammonia industry. They depend on the characteristics of reservoir rocks, gas composition and fluid mechanics. Good quality rocks of large permeability, few thousand m depth... are being exhausted and so the consumption rates for the gas production are increasing. Transport distances and subsequently energy losses are also increasing, although the use of higher gas pressure in pipelines works against this trend. The conversions to ammonia and then to fertilizers itself are reaching a theoretical limit as seen above. The losses of N from the fertilizer to the plant are variable but in average about 20% [12], which are already relatively low. Nitrogen is contained in the plant largely in form of proteins (but not limited to).

The "efficiency" of the rest of the chain, from the plants to human food *via* animal proteins or directly, is more related to the consumer behavior (see section 3.2).

**From Steam Engine to Turbine in Power Station** Section 5.1.1 of the chapter 5 shows a similar trend in efficiency for the steam engine. A gain larger than two hundred fold was obtained from the Newcomen machine in 1712 to the modern steam turbines in large power plants. Future progress in conversion of heat into work can only be much smaller according to fundamental and engineering knowledge.

**Steel Production and Their Use** As a cautious note about the limitations on future gains we can also look at the progress made in steel manufacture. The chain of transformations covers operations from iron extraction or scrap collection to manufacture of equipments such as power generators (nuclear plants or wind turbines) and ships.

Similar to ammonia production, steel production based on reduction of iron oxide in modern blast furnaces followed by the iron pig conversion in basic oxygen furnaces has probably reached its limits in terms of energy efficiency. The technology is responsible for about 70% of the global steel production, which amounted to 1 500 millions tonnes per year in 2011 according to the World steel association. The whole industry was largely improved and developed at a large scale during the industrial revolution. It benefited from accumulated knowledge over 500 years as well as the availability of large quantity of coal. High and homogeneous temperature is reached over a large volume of blast furnace - with current capacity larger than 1 Mt per year - along with the improved characteristics of inputs, ore and reducing agents (like coke). Integration including in one site the furnaces as well as auxiliaries such as coke ovens, ore pellet plants and steel product forming have facilitated energy savings [13]. The International Energy Agency IEA reports that the American blast furnace industry reduced its consumption of final energies - mostly coal LHV - from 21.2 $MJ_{LHV}{\cdot}kg_{stl}^{-1}$ in 2002 to 20.3 $MJ_{LHV}{\cdot}kg_{stl}^{-1}$ in 2005. A part resulted from closure of old sites (3 out of a total of 19) and increased use of scraps instead of iron ore (according to information from the American Iron and Steel institute). Remaining gains of efficiency rests on better recovery of sensible heats from outputs like slag and steel of basic oxygen furnace, or of work from high pressure off-gas, which amount to a few percent of overall consumption [14].

Other process routes such as steel from electric arc furnaces using iron scraps - about 25% of the global steel production - or direct reduction of ore without melting - about 5% of the production - present better energy efficiency, even expressed in primary energy. The IEA reports that the American electric arc based industry reduced its consumption of final energies - largely electricity - from 5.2 $MJ_{LHV}\cdot kg_{stl}^{-1}$ in 2002 to 4.9 $MJ_{LHV}\cdot kg_{stl}^{-1}$ in 2005 (the industry benefited from the start of new electric furnaces). In addition these furnaces are much smaller and more flexible than the blast ones. However, they depend on very high quality and limited feedstock. Further gains for these technologies are also small [14].

A more promising technology to substitute for the blast furnace/oxygen converter industry is the smelting reduction of ore without using coke. Ore is first reduced with a synthesis gas produced from coal gasification and partial oxidation. Iron is further reduced and melted in the reactor where coal gasification takes place. However, coal consumption of this technology is still high and gains of energy efficiency depends on the use of the exhaust gas [14].

On the other hand, the industry is able to make high quality - such as high strength or low fatigue - or specialized steels owing to addition of other elements (chrome, nickel, molybdenum, vanadium...) and/or a better refining of raw steel and an accurate control of thermal and mechanical treatments for finished steel products. In addition, by design and final transformation, downstream industries allow these new products to reduce the amount of steel for the same application (mass per unit of service over the equipment lifetime expressed by a rate $R_D$ as defined in chapter 2). An example is provided in section 3.3.2 with nuclear power plants. The gains are measured by a rate expressed in mass of steel per produced kWh over the plant lifetime.

### 3.1.2. Recent Trends in Energy Intensity and Efficiency of the US Economy

Rather than the energy consumption per unit of physical output, $R_D$ or $R_{EP}$, an indicator used by economists to measure the efficiency of a system - or economic unit - $M$ is its energy consumption per unit of its added value, the energy intensity $I_D$.

The added value is the difference between the costs $\$_D$ of energy and materials, purchased from other economic units for the operations and investments of the system $M$, and its sales $\$_{out}$ [15]. The added value is redistributed as salary of workers, benefits paid to investors and taxes raised by the government to the industry $M$. A commonly used quantity in the macro-economy, the gross domestic product GDP of a country, corresponds to the aggregate added values generated by all economic units of the country over a fixed period, usually a year [15]. Added values, GDP and the monetary flows to determine them are gathered and treated by government agencies, such as the US census bureau, for the country national accounts. To achieve this task all the various activities are split in different sectors, subsectors, groups of industries and industries according to the type of their production and a standardized classification such as the North American Industry Classification System NAICS (agriculture, mining, manufacturing...). Government as a service provider is part of the economic activity (even if not directly paid). On the other hand the households of the country and their own direct production (such as the vegetables from their garden for their own or friend's use) are excluded, as they only represent end users of the goods and services of the economic activities.

The intensity $I_D$ of an economic unit $M$ is not deduced from the sale of its output, or revenue, because part of it is used to pay for the goods and services purchased from other units. The manufacture of these intermediate products have required some energy, which should be included in the energy consumption in the intensity $I_D$ if the revenue is chosen for its denominator, like in the case of the rate $R_D$. Now, the intensity takes only into account the direct consumptions of final energy - electricity and processed fuels - by the unit $M$. Hence the use of the added value for the calculation of $I_D$.

This section examines various economic units in the USA and their inputs and outputs for the years 2002 and 2007. The economic units investigated are the whole economy of the country, the subsector of machine manufacturing (NAICS code 333) - one subsector of the manufacturing sector (NAICS code 31-33) -, and the petroleum refineries (NAICS code 32411) - an industry of the group of industries *petroleum and coal manufacturing* (NAICS code 3241) -.

The ammonia industry, albeit not part of the economic classification, is also studied.

The energy intensity $I_D$ as well as the monetary share of final energy in the added value are determined for each unit and each year from their monetary and energy flows. This work permits a comparison between different units of each indicator, as well as their evolution over the span of five years. This period takes place before the major economic crises starting after the end of 2008 and, consequently, the units are not affected by the crises. However, the period is associated to a rapid rise of energy price, especially for crude oil.

At the difference of monetary flows, physical flows for an economic unit, in particular its output $L_{out}$, are difficult to retrieve. The US census bureau reports only the electricity purchase of each economic unit. Energy intensive sectors like petroleum refinery are, however, documented thanks to other sources such as the US department of Energy. As a result the rate $R_D$ is only known for the petroleum refinery and ammonia industry. It is tentatively determined for the machinery manufacture from the steel inputs of the sector.

Over the short span of time of five years the improvements on the rate $R_D$ are usually not very significant (see Fig. 3.1.1 for recent years). It may be important when foreclosure of old and inefficient factories of the sector under study, or/and starting of modern and more efficient ones represent a large share of the operating capacity, as it seems the case of the photovoltaic industry (section 3.1.3).

### At the Level of the Country: GDP

In Fig. 3.1.2 the energy flows consumed by the USA are shown for 2002 and 2007, according to data compiled annually by the US Energy Information Administration for its Annual Energy Review [16].

The energy flows from natural resources - or primary energies (petroleum, coal...) - on the left to the end users (industry, commerce, transport, household) on the right, with the conversion into fuels and electricity - or final energies - on the middle. The flows includes the different energy imported; in terms of heat value, 66% of petroleum and 16.5% of natural gas were imported in 2007, while in comparison importation in 2002 amounted to 60% of petroleum and 16% of natural gas (these amounts have decreased since).

Energy forms are measured with their heat value expressed in Mtoe, using the American convention - fuels in HHV -, and, in parenthesis, with their monetary value expressed in $ of the year (or nominal $). The values are rounded to take margins of error into account, even if original data were given with higher digit accuracy. It has been checked that the sum of the energy supplied by the sources equals the sum of the energy used by the various consumers, to make sure the law of energy conservation is maintained.

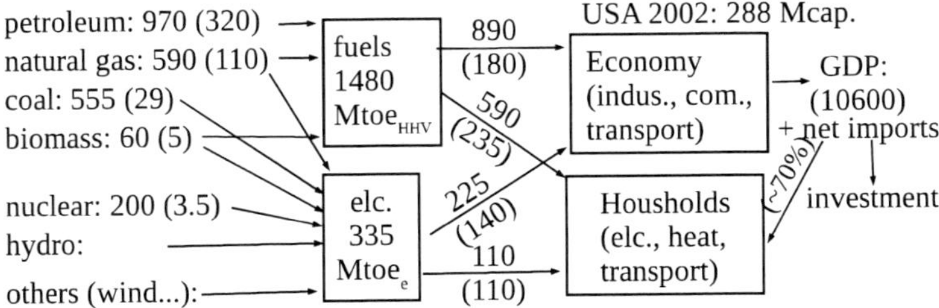


(a) Energy and monetary flows for the USA in 2002.

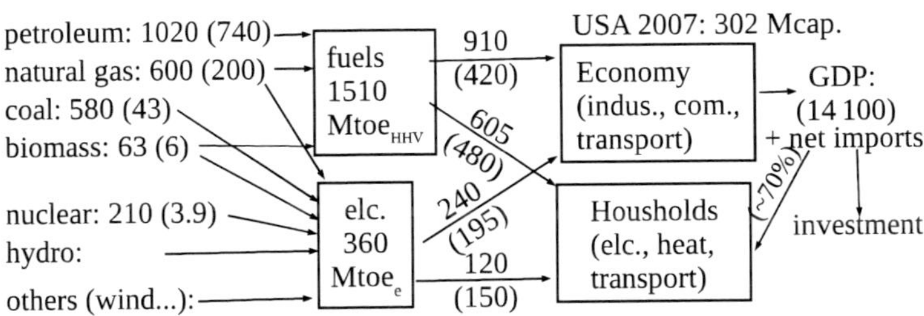


(b) Energy and monetary flows for the USA in 2007.

Figure 3.1.2. Energy flows in 2002 and 2007 for the US economy and households, expressed in $\mathrm{Mtoe_{HHV}}$ for fuels and in $\mathrm{Mtoe_e}$ for electricity (toe: tonne oil equivalent). Data in parenthesis correspond to the associated monetary flows in nominal G$ (billion dollars). Source: US Energy Information Administration (USEIA) for the Annual Energy Review [16]. *cap* represents the number of US inhabitants at the end of each year.

Since private residential households sector is not part of the productive system, their fuel and electricity consumptions are not included in the economic requirements. Due to the characteristics of the transport sector in the USA, the households are assumed to consume exclusively and entirely the supply of motor gasoline (including agro-ethanol). All the other petroleum products are consumed for the US economy.

On the other hand, all consumption occurring in the energy transport and refining sectors is included in the economy, except the important loss of conversion into electricity.

The monetary value of the final energy represented 3.0% of the GDP in 2002. It rose to 4.4% in 2007. The share of imports, mostly petroleum and gas, amounted to about 2.0% of the GDP in 2002 and 3.5% in 2007. The increase resulted from the rise of energy prices between the two years, especially petroleum one, nearly two and half fold, and gas one nearly twofold. In spite of these price spikes the consumption of each energy was higher in 2007 than in 2002, showing how vital energy is. The economy and households were ready and able to pay the elevated price, at least until 2007.

In terms of energy intensity, $I_D$, 0.105 $toe_{EF}$ of final energies were required per k$ of GDP for 2002, while $I_D$ amounted to 0.082 $toe_{EF}$ per k$ for 2007. The main reason of the reduction of the intensity was the growth of the GDP - about 32% between 2002 and 2007 -, as the energy consumption increased only by 2% for fuels and by 8% for electricity.

Part of the GDP rise resulted from general inflation. It is corrected with a technique of the US administration called the chained-dollar measure. Nevertheless it implies some arbitrary choices to separate price variation of a product due to average monetary inflation - the real inflation - from the price increase due to new functionality or product improvements - a real contribution of added value that justifies some of the price increases[1].

From the corrections given in ref. [16], the GDP expressed in 2005 $ were 11 500 and 13 200 $G\$_{05}$ for 2002 and 2007 respectively. The growth of the corrected GDP was close to 15% (or 2.7% per year), reducing the difference between the energy intensities (0.097 and 0.087 $toe_{EF}$ of final energies per $k\$_{05}$ for 2002 and 2007 respectively). The increase of energy consumption remained lower than the GDP growth. For fuels the growth was close to the population growth between 2002 and 2007 - +1.5% -. The larger increase of electricity consumption is probably a consequence of the high prices of petroleum and natural gas (electricity prices increased far less).

In summary the increase of the monetary share of the energy in GDP resulted mainly from the large price rise between 2002 and 2007, and not a reduction of the efficiency, which would otherwise be worrisome.

As far as the intensity $I_D$ is concerned, its diminution resulted largely from

[1]Recent price spikes of energy are thus part of true inflation. Inflation is thus not homogeneous and its correction at the level of an economic unit with factor determined at the national level is inadequate. The US bureau of labor statistics provides price indexes of main products by year. However, the detailed physical inputs and outputs of each economic unit, required to correct for inflation with indexes, are not often available.

the GDP growth, even corrected for inflation.

The same type of study is performed at the level of a subsector.

### At the Level of a Subsector: Machinery Manufacturing

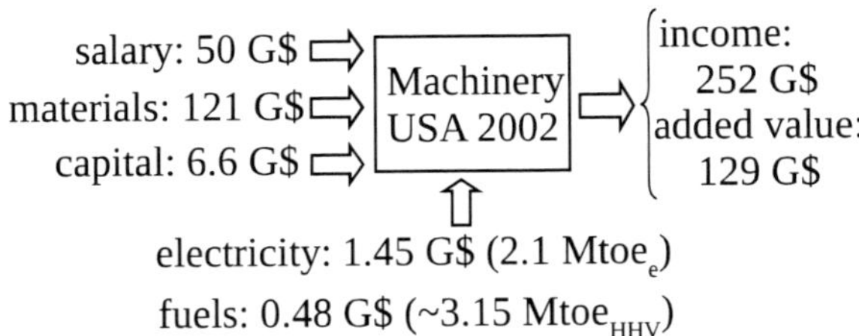


(a) Flows for the US machinery manufacturing in 2002

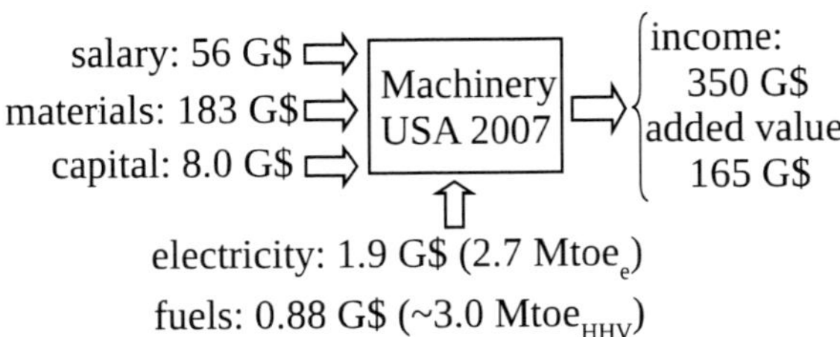


(b) Flows for the US machinery manufacturing in 2007

Figure 3.1.3. Monetary inputs and outputs for the US machinery manufacturing. Source: US economic census 2002 and 2007 code NAICS: 333 [17]. The energy content of fuels - assumed only gas - was derived from the industrial price of gas reported by the US Energy Information Administration [18].

The subsector comprises different industries, output of each of them aimed at a particular sector of the American economy from the agriculture and mining to the automotive and clothing industries. They produce machine tools, construction and mining equipment, processing, handling and packaging machinery, turbines, electric generators and power transmission equipment.

Fig. 3.1.3 shows the monetary flows of the subsector for the USA in 2002 and 2007, according to the US census bureau [17]. The agency indicates the amount of electricity purchased and its cost during the year. Only the cost of the fuel consumption is reported. By assuming gas the only fuel purchased and by using the industrial average price of the gas for each year, the expenditure is converted into an energy flow. The gas price rose twofold from 2002 to 2007 (153 $\$.toe_{HHV}^{-1}$ against 293 $\$.toe_{HHV}^{-1}$), but the consumption remained similar.

For 2007 the monetary value of final energy represented 1.7% of the added value of this sector (in 2002 it was 1.4% of the added value). Like for the whole US economy the energy costs of this economic unit are negligible relative to its added value.

In terms of energy intensity $I_D$, an average of 0.034 $toe_{EF}$ per nominal k$ of added value was required by this subsector for 2007, while it amounted to 0.047 $toe_{EF}$ per k$ for 2002. As the energy consumption rose to around 8% - only due to electricity -, the variation resulted mostly from that of the added value, 29% in nominal $. There is no information to correct for the inflation specific to this economic unit. If we take the average inflation for the whole US economy, the increase of added value would amount to about 14%, lessening the reduction of $I_D$.

We can estimate the rate $R_D$ of direct consumptions for this subsector from its final energy consumptions and from one of its main inputs, finished steel products supplied by the iron and steel industry. Like for the outputs the US census bureau does not provide any physical quantity about purchased materials. However, according to information from the American Iron and Steel Institute, the machinery subsector bought 1.27 Mt worth of steel for both 2002 and 2007. According to the US census bureau the number of establishments had diminished by 13% while the number of employees has remained unchanged (firms reduction seems inconsistent). Hence it is reasonable to assume that the output has not increased much and, consequently, the consumption rate $R_D$ - including steel or not - had not diminished at the same pace as for the intensity $I_D$, if it diminished at all.

We can characterize this subsector as a high added value economic unit with low monetary dependence on energy. It will be much more affected by the market prices of its products and of its purchase of non energy related feedstock than by the energy price.

### At the Level of an Energy Industry: Petroleum Refinery

In Fig. 3.1.4 the monetary flows for the American petroleum refinery industry in 2002 (Fig. 3.1.4(a)) and 2007 (Fig. 3.1.4(b)) are shown based on data of the US census bureau for the economic census of 2002 and 2007 [17].

The US bureau information for material and fuels consumed in 2007 by the

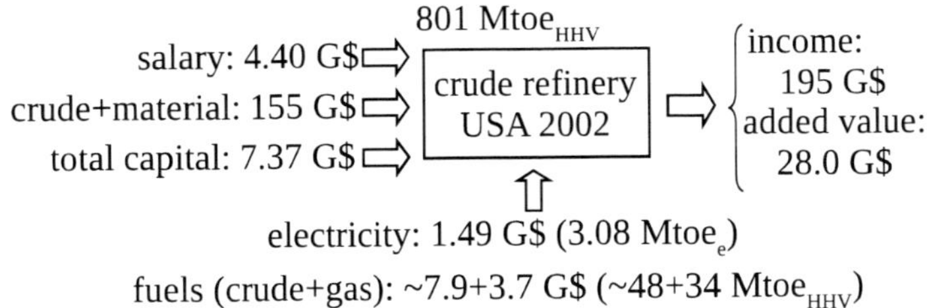


(a) Flows for the US petroleum refineries in 2002

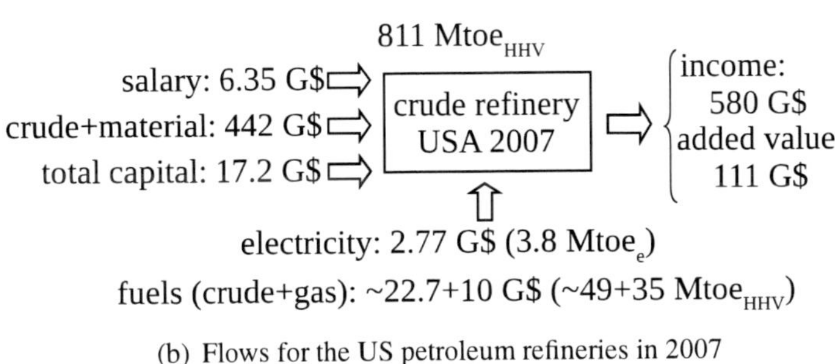


(b) Flows for the US petroleum refineries in 2007

Figure 3.1.4. Monetary inputs and outputs of the US petroleum refineries in 2002 (a) and 2007 (b). Source: US economic census code NAICS: 32411 [17]. Monetary and heat value flows for the energy consumptions were deduced from another source: the US Energy Information Administration [18, 19]. The cost of the energy purchases was subtracted from the feedstock and material costs.

industry indicates the purchase of 23 Gb - b = barrel, about 0.14 toe of heat value - of domestic crude oil, whereas crude import amounted to only 3 Gb. The list also includes the inputs of 16.3 Gb of toluene and xylene, which represent rather chemical outputs of a refinery. The consumption of chemicals like catalysts is not mentioned. Physical quantities and types of consumptions appear thus erroneous (not to say fanciful). On the other hand the delivered costs (in $) are more plausible. The comparison with the data from the US Energy Information Administration USEIA [19] shows a discrepancy for the fuels, while the data are close for electricity.

Albeit tedious due to the conversion from barrels to heat values, the energy balance of the US refinery can be obtained from the USEIA data [20]. The average refinery consumes for its operations about 6% of its crude input and a quantity of natural gas equivalent to 4.3% of the crude input (used to produce heat, $H_2$ and electricity). These rates of consumption appear constant over the years and are used for both 2002 and 2007. Using the industrial prices of crude and gas from the USEIA for each year (for crude oil: 463 $\$.toe_{HHV}^{-1}$ in 2007 and 164 $\$.toe_{HHV}^{-1}$ in 2002), the monetary values of the fuels consumed are deduced and reported in Fig. 3.1.4. The expenditures of gas were close to the overall values reported in the economic census for fuels (4.3 G$ and 8.6 G$ for 2002 and 2007 respectively). The crude consumed in the refineries to fuel their processes was probably not taken into account.

For 2007 the monetary value of the energy consumptions by the sector was equivalent to 32% of its added value (47% in 2002).

In terms of energy intensity, an average of 0.79 $toe_{EF}$ per nominal k$ of added value is required by this industry for 2007, while $I_D$ was as high as 3.0 $toe_{EF}$ per k$ for 2002.

The economic and energy pictures of refineries are very different from the country economy and the subsector looked at above. The cost of the energy consumed is of the same order of the added value, albeit the latter increased faster than the former from 2002 to 2007. From the data it appears that the increase of fuel expenditure was partly offset by the purchase of electricity and gas, the prices of which rose more slowly than the crude oil prices. The industry feedstock and its products, namely crude oil and petroleum products, are also energies, and their prices have experienced large increase with fluctuations since 2002. As the difference of two fluctuating quantities, the added value presented large fluctuations too, with an overall increase between 2002 and 2007 (thereafter the added value has decreased with a low at 68 G$ in 2009).

On the other hand, in terms of heat values, the energy input and the energy consumption remained fairly constant, experiencing an increase of less than 1.5%. As a result, the energy intensity $I_D$ of the refinery industry varied inversely with its added value. In all cases the intensity $I_D$ remained very strong in comparison with the US economy one. Based on its change it shows little relation to the rate of consumption $R_D$.

### At the Level of an Energy Industry: The Ammonia Producers

Table 3.1.1 shows the monetary flows (sales, costs and margins) of the ammonia $NH_3$ industry in the USA for 1995, 2002 and 2007. In the standardized

**Table 3.1.1. Sales [21] and gas costs for the American $NH_3$ industry for 1995, 2002 and 2007 in nominal \$. Gas costs are deduced from the industrial price of gas [18] and the average rate $R_D^{NH_3}$ for the USA in 2007 (38 $GJ_{HHV}.t_{NH_3}^{-1}$ [21]). The industry margin for a year is approximated to the difference between its sale and gas purchase (gas represents its major purchase). The last column shows the production of $NH_3$ in the USA according to the US Geological Survey minerals information**

| Year | sale $\$.t_{NH_3}^{-1}$ | gas cost $\$.GJ_{HHV}^{-1}$ | gas price $\$.t_{NH_3}^{-1}$ | margin $\$.t_{NH_3}^{-1}$ | gas cost % margin | prod. $Mt_{NH_3}$ |
|---|---|---|---|---|---|---|
| 1995 | ~ 235 | ~ 1.5 | ~ 57 | ~ 180 | ~ 32% | 15.8 |
| 2002 | ~ 145 | ~ 3.7 | ~ 140 | ~ 5 | ~2 800% | 12.5 |
| 2007 | ~ 360 | ~ 7.05 | ~ 270 | ~ 90 | ~ 300% | 10.4 |

classification of the economy the industry is included in the N fertilizer industry (NAICS code 325311). Consequently no information specific to $NH_3$ industry can be derived from the economic census. Data are provided by a note of the Economic Research Service within the U.S. Department of Agriculture [21] and by the series of the USEIA about the industrial price of gas [18].

The cost of the gas input is expressed in $\$.t_{NH_3}^{-1}$ owing to the rate $R_D^{NH_3}$ of the industry in the USA (38 $GJ_{HHV}\cdot t_{NH_3}^{-1}$ [21]; see also the section 3.1.1).

An industry margin is calculated from the difference between sale and gas cost. It is approximated to the added value of this industry. The cost of the gas as a fraction of the margin varied between 32% in 1995 and 2 800% in 2002. The variations of the gas and the ammonia prices did not simultaneously cancel each other out.

The energy intensity $I_D$, deduced from the rate $R_D^{NH_3}$ (0.91 $toe_{HHV}\cdot t_{NH_3}^{-1}$) and the financial margin, amounted to about 5.0, 180 and 10.0 $toe_{HHV}$ per nominal k\$ of added value for 1995, 2002 and 2007 respectively. This very energy intensive industry suffered from high gas prices between 2000 and 2008, reducing its margin and leading to the foreclosure of factories (last column of table 3.1.1), the oldest and less efficient ones. Due to theses foreclosures the US average rate $R_D^{NH_3}$ has probably diminished during that period. However, because of limited gains for recent factories - as seen in the section 3.1.1 -, the variation has remained smaller than that of intensity $I_D$.

Like in the case of the refinery industry - and with even more striking data -, the relationship between the energy efficiency measured with the rate $R_D^{NH_3}$ and the energy intensity $I_D$ has weakened. Both indicators acted even independently during these two last decades, with the intensity controlled by the sale prices and the cost of feedstock. Before, the rate $R_D^{NH_3}$ had probably played the driving role in relation to $I_D$, as evidenced by Fig. 3.1.1.

The foreclosures resulted in an increase of ammonia imports. This situation would represent a form of 'economic logic' if the employees laid off would have been able to find a work in another sectors. There is a large chance that their new sectors produce higher added value items than ammonia and with at least the same labor productivity. Thus the reduction of added value due to the decrease of the ammonia production could be largely offset. The overall energy intensity of the country economy is lowered, as a result.

The global rate $R_D^{NH_3}$ can also improve, but independent of the variation of $I_D$ in the USA. Indeed the ammonia exporters are probably natural gas producers developing a chemical sector with modern factories (like in Trinidad and Tobago). Moreover, the transport of ammonia in liquid form requires less fuel than the transport of gas by pipeline or by ship in liquid form at low temperature.

In 2011 the American production remained fairly constant at 9.8 $\mathrm{Mt_{NH_3}}$ according to the US Geological Survey minerals information. However, the financial situation has dramatically improved. The average sale price of ammonia rose to 570 $\$.\mathrm{t_{NH_3}^{-1}}$ whereas natural gas cost decreased due to the 2008 crisis and a reduction in its demand (price of 4.0 $\$.\mathrm{GJ_{HHV}^{-1}}$ on average for 2011 at the Henry Hub in Texas). Cost also reduced as a result of the production of gas from very tight rocks and even source rocks, which permitted to largely offset the decline of production from other sources. The cost of the gas as a fraction of the margin was 36% in 2011 like for 1995. The intensity $I_D$ fell to about 2.2 $\mathrm{toe_{HHV}}$ per nominal k$ of margin, assuming the same rate $R_D^{NH_3}$.

This improvement was largely obtained at the expense of the consumers and therefore the whole economy.

## Some Remarks on the Energy Indicators from the Study of the American Economy

The monetary and heat values, respectively expressed in $ and $\mathrm{toe_{EF}}$, of the energy consumed by the overall US economic activity, one of its subsectors, and two of its industries have been reported and studied per unit of aggregate added value of each economic unit. Based on this work the economic units can be separated broadly into two categories:

- For the vast majority of the units, the final energy purchases represent a small fraction of the added value - a few percent - even with the large increase of the energy price as recently experienced. They also show a low energy intensity $I_D$.
- Industries usually involved in the transformation of energy or raw material require energy themselves for operation in the form of fuel and electricity purchases representing more than 10% of their added value. As a consequence they are intensive energy industries.

What about the rate $R_D$ of these two types of economic units?

**High Added Value Units** The economic units of the first category can be defined as high added value units. The inverse of their energy intensity is maybe of greater significance: for each unit of heat value consumed there is a corresponding amount of added value, which can be canceled if this heat value is not available. We can think about the service sectors, such as the tourism, which depend directly or indirectly on the transportation of passengers and freight, and so on its fuel requirements. In the short and medium terms they are able to absorb an increase of the energy price.

The financial surveys do not provide the raw data to derive an estimate of the rate $R_D$ for these economic units. However, because their fuel expenditures represent a small share of their added value (and of their revenue), there is little incentive to improve efficiency in these sectors since it would require an increase of capital expenditure.

In the example shown above, the machinery manufacturing, the consumption of final energy has increased from 2002 to 2007, albeit only by 10%, in spite of higher prices. Yet energy intensity $I_D$ decreased during the same period owing to an increase of the revenue by 40% and in spite of the rise of material costs by 30%, among which steel (in nominal values). The information about physical outputs is not available. However, the input of steel mill products - the main material in physical term - remained unchanged. The capital expenditure grew by a mere 20% from 2002 to 2007, close to the official inflation, which is not an indication of large investment to improve energy efficiency. As result the increase of the revenue resulted probably from an increase of the prices of the products.

**Low Added Value Units** The rates of consumption $R_D$ of the industries given in the examples - petroleum refineries and ammonia industries in the USA - were derived from other sources than the economic surveys. At least for the refinery sector since 2002, a reduction of $R_D$ was not observed in spite of a larger incentive than for the economic units of the first category to achieve it due to the rise of the energy cost.

On the other hand, large variations of the share of the energy cost in the financial balance, and of the intensity $I_D$, were observed in the last 15 years. The main inputs of these industries as well as their outputs are some forms of energy. Hence the large variations of their monetary indicators resulted solely from the fluctuations of the prices of the different forms of energy.

**Overall Economy** The values of the monetary indicators of energy consumptions, such as the energy cost as a fraction of added value and energy intensity, for the country wide economy are similar to the ones for the high added value units, as observed recently in the USA. The reduction of the rate $R_D$ in energy and raw material industries, as well as the physical gains in their labor and investment productivities, over the long period of the industrial revolution had been responsible for their present low monetary weight in the overall economy, at least for developed countries. The same evolution also happened for agriculture and production of food, the fuel and building elements of human being.

But how small can be the weight of energy and raw material industries in the economy due to the relatively low price of their products?

The economy presents us with a paradox in this case: energy in heat value has become vital for a modern economy, whereas its monetary value has suffered devaluation. If the physical rate $R_D$ can not decrease further, and even increase due to poorer feedstock (necessary exploitation of new reserves of lower quality than previously), the situation can lead to falling incomes for producers. A crisis of the sector and a reduction of the number of producers can ensue. Because of the possible formation of monopolies and the limitation in the rate $R_D$, a rise of energy prices becomes inevitable, as observed for ammonia and crude oil prices.

### 3.1.3. Variations of Prices and Their Origins: Case of the Photovoltaic Sector

The price changes of a product, and subsequently its added value for the suppliers, can result from an artificial demand driven by government incentives and/or from erroneous market expectations. An example is the worldwide photovoltaic PV market and its development from 2000 to 2012.

The annual manufacture of PV cells has increased from about 0.25 $GW_p$ in 2001 to about 35 $GW_p$ in 2011, according to the International Energy Agency ($W_p$ is the unit of peak or nominal power of the module made of the cells. See technical definition below). This dramatic rise was motivated by a strong demand itself spurred by public incentives such as guaranteed grid feed-in tariff and tax credits for producers of PV electricity, as looked at in depth in the following. As far as the PV industry is concerned, in a few years it has moved from a situation of scarcity of materials to satisfy the demand, and of high prices of its products, to a situation of overcapacity and depressed prices.

#### Indicators of Efficiency and Profitability of the Photovoltaic Sector

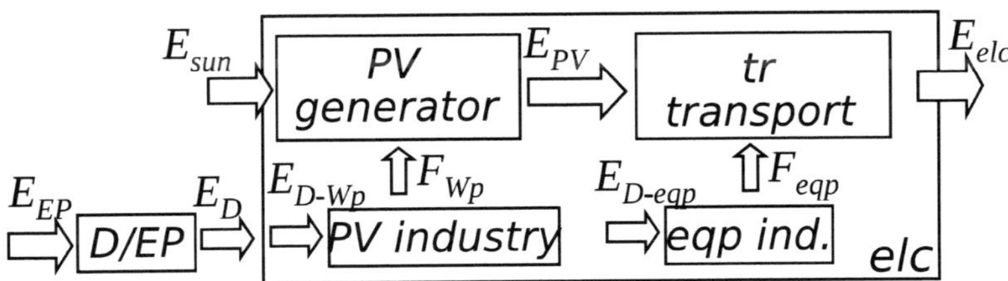


Figure 3.1.5. Schematic diagram of flows of electricity production *elc* from solar input $E_{sun}$ to the final output $E_{elc}$ at user level. To produce intermediate flow $E_{PV}$ at the PV generator *PV*, the PV industry manufactures $F_{W_p}$ modules while consuming $E_{D-W_p}$ of direct requirements. Likewise $F_{eqp}$ equipment are required to perform transport and regulation *tr* of electricity. Total consumption $E_D$ by the system *elc* requires the primary energy input $E_{EP}$.

**Energy and Material Flows** The system *elc* of electricity production from the sun and its subsystems (PV generator, PV industry...) are modeled by their flows of energy and materials over a fixed period such as the lifetime of PV modules (Fig. 3.1.5):

- The sun input $E_{sun}$ on PV generator $PV$ in kWh$_{\text{ph}}$ of radiation, the corresponding output $E_{PV}$ at PV generator very outlet in kWh$_{\text{PV}}$ of direct current DC electricity and output $E_{elc}$ at user socket in kWh$_{\text{elc}}$ of alternative current AC electricity.

- The flow $F_{W_p}$ of PV modules in W$_{\text{p}}$ to produce $E_{PV}$, and the flow $F_{eqp}$ of other equipment required in operation $tr$ to convert $E_{PV}$ into $E_{elc}$.

- The direct requirements $E_{D-W_p}$, $E_{D-eqp}$ or their sum $E_D$, as well as associated primary energy consumptions $E_{EP-W_p}$ or $E_{EP}$, for the whole system $elc$ and its subsystems to produce $E_{elc}$.

The PV industry includes operations from mineral extraction to module encapsulation *via* semi-conductor purifications. Its requirements $E_{D-W_p}$ represent the consumptions of electricity, coal, chemicals, machinery... They coincide with the main consumptions $E_{D-PV}$ for PV generator as inverter and transformer necessary to adapt the electricity to the specifications of grid or household appliances are included in the operation $tr$ ($E_{D-PV} \equiv E_{D-W_p}$).

The flow $F_{eqp}$ also comprises the network equipment between generator and user socket. It may also include the consumptions to make batteries in the case of an off-grid site.

**Consumptions Rates** The rate $R^{elc}$ of consumptions, either in direct or primary requirements, for the whole system is:

$$R^{elc} = (E_D \,\text{or}\, E_{EP})/E_{elc}.$$

It is made of the consumption rates of its two main operations, PV generation $PV$ and transport $tr$, $R^{elc} = R^{elc}_{PV} + R^{elc}_{tr}$.

The subsystem $PV$ can be studied independently using its rate:

$$R^{PV} = (E_{D-PV} \,\text{or}\, E_{EP-PV})/E_{PV}.$$

Likewise the PV industry is looked at in isolation with its own rate:

$$R^{W_p} = (E_{D-W_p} \,\text{or}\, E_{EP-W_p})/F_{W_p}.$$

Specific rates to operations $tr$ and $eqp$ can be also defined ($R^{tr}$ and $R^{eqp}$).

We have the following relations between above rates:

$$\begin{aligned} R^{elc}_{PV} &= R^{PV} \cdot w^{elc}_{PV}, \\ R^{PV} &= R^{W_p} \cdot w^{elc}_{PV}, \end{aligned}$$

where

$$w^{elc}_{PV} = \frac{E_{PV}}{E_{elc}} \quad \text{and} \quad w^{PV}_{W_p} = \frac{F_{W_p}}{E_{PV}}.$$

**Meaning and Values of Quantities $w^{elc}_{PV}$ and $w^{PV}_{W_p}$** The term $w^{elc}_{PV}$ represents the total amount of electricity $E_{PV}$ required to offset the dissipations in the subsystem $tr$ and obtain one $\text{kWh}_{\text{elc}}$ at the user socket. The value depends on whether the PV modules operate with unlimited connection to a national grid, or are off grid using batteries to store electricity for adjustment to the demand. Both situations include the dissipation in inverter and transformer, about 5%.

The losses in the off-grid case result from the chemical conversions in batteries and from the disconnection of PV module from the batteries when their capacity is reached. In the case of a household the losses amount to about 45% of module output when producing without restriction. Hence the term $w^{elc}_{PV}$ amounts to $w^{elc}_{PV} = 1.8\ \text{kWh}_{\text{PV}}{\cdot}\text{kWh}^{-1}_{\text{elc}}$.

When connected to a country wide grid without restriction the loss corresponds to the electricity dissipated in the network. In France it represents about 7% of the total net output of all power generators. Including inverter and transformer dissipations the total amounts to about 10%. The term $w^{elc}_{PV}$ is thus $w^{elc}_{PV} = 1.1\ \text{kWh}_{\text{PV}}{\cdot}\text{kWh}_{\text{elc}}$.

Most of the PV installations since 2001 have been grid connected systems due to the various national incentives. The contribution of off-grid modules is thus discarded for this study. However, the regulation contribution should not be forgotten. In the extreme case where PV generators represent the main source of electricity (potentially on an island), the grid has to be managed like in the off grid situation with consequently much larger losses than currently incurred.

The inverse of the quantity $w^{PV}_{W_p}$ represents the amount of DC electricity produced at the outlet of 1 $\text{W}_\text{p}$ of PV module during its lifetime $T_{W_p}$. By also introducing the surface $S_{W_p}$ of $F_{W_p}$ modules, the inverse of $w^{PV}_{W_p}$ is expressed by the product of three separate terms:

$$\left(w^{PV}_{W_p}\right)^{-1} = \left(\frac{S_{W_p}}{F_{W_p}}\right) \cdot \left(\frac{E_{sun}}{S_{W_p} \cdot T_{W_p}}\right) \cdot \left(\frac{E_{PV}}{E_{sun}} \cdot T_{W_p}\right)$$

The inverse of the first term, $\eta_{W_p} = F_{W_p}/S_{W_p}$ in $W_p$ per square meter of module or $W_p \cdot m_{PV}^{-2}$, is measured at the factory before shipment and is provided by makers in the specification sheet of the module (see the next paragraph about PV industry for values). It is the DC power produced at module temperature of 25°C under a sun radiation of $E_{sun} = 1\,000$ $W_{ph} \cdot m_{PV}^{-2}$ (close to the maximal sun radiation on Earth surface). Owing to its definition the factor $\eta_{W_p}$ is formally equivalent to a module yield expressed in $W_p$ per $W_{ph}$.

The second term in $W_{ph} \cdot m_{PV}^{-2}$ is the sun power on the module averaged over its lifetime. It depends on the local sun irradiance, averaged over sufficient time, and on the module orientation - usually at fixed angle according to the local latitude -. It takes into account, or should, natural shadings such as that due to clouds. Recent production in Spain is about 22% of connected $W_p$ whereas the yield is close to 11% in Germany (from data of average capacity and production for a year in each country provided by government agencies). Because the difference results only from irradiance conditions - similar modules and optimal orientations -, and using the definition of module peak power, the average irradiance in Spain was about 0.22 $kW_{ph} \cdot m_{PV}^{-2}$ and two times as low in Germany.

The third term is the module yield integrated over its lifetime. The yield $Y_{W_p} = E_{PV}/E_{sun}$ in the first years of operation can be approximated by the term $\eta_{W_p}$ in $W_p \cdot W_{ph}^{-1}$ ($Y_{W_p}$ should actually include module mismatch, as the generator is usually made of an assembly of modules, and dust shading). However, the yield degrades over time. It results from exposition to natural elements such as salinity of rainfalls, moisture and aerosols of air, high temperature and ultraviolet UV radiations. Manufacturers usually guarantee that a module retains at least 70% of its initial yield after 25 year operation. As most of present capacity is less than ten year old, these guarantees are based on the behavior of old modules made with the same technology (crystalline Si one exists since 1970's) and accelerated ageing tests in laboratory. On the other hand, the premature breakdown of part of the modules due to manufacturing defects, which have intensified recently [22], must be included. From large sampling - few hundred of thousand modules - the defect rate has reached 13%. Destruction before 25 years can also occur due to natural or human made events.

Following PV maker indications the term $Y_{W_p} \cdot T_{W_p}$ can be approximated by $\eta_{W_p} \cdot T_{W_p}$ with $T_{W_p}$ = 25 years or 220 kh. Hence the inverse of $w_{W_p}^{PV}$ in

$kWh_{PV} \cdot W_p^{-1}$ is close to:

$$\left(w_{W_p}^{PV}\right)^{-1} = 220 \cdot \frac{E_{sun}}{S_{W_p} \cdot T_{W_p}} (\text{in } kW_{ph} \cdot m_{PV}^{-2})$$

.

The term would amount to 48 $kWh_{PV} \cdot W_p^{-1}$ in Spain while 24 $kWh_{PV} \cdot W_p^{-1}$ in Germany.

Our model assumes that the lifetime is independent of the average solar irradiance. It means that degradation results mainly from moisture and/or aerosol actions. In the case of degradation resulting from solar irradiance, such as the amount of UV, lifetime can diminish with rising irradiance, resulting in similar total input $E_{sun}$ over module lifetime - and consequently similar output $E_{PV}$ - whatever the place. In absence of more information we assume the validity of the first model, which shows a clear advantage for sites with high irradiance.

**Monetary Flows and Added Values** Similar to energy and material flows, we can define monetary flows for the system *elc* over the same period (such as lifetime of PV module):

- The revenues from electricity sales at the socket, $\$_{elc}$, and at the PV generator, $\$_{PV}$.
- The revenue $\$_{W_p}$ of PV industry from the sale of $F_{W_p}$ modules.
- The costs of energy and materials for PV manufacturers, $\$_{D-W_p}$, and for the whole system, $\$_D$. The costs of equipment $F_{eqp}$ could be included owing to $\$_{eqp}$. On the other hand sun is free ($\$_{sun} = 0$).

The investor in PV modules pays the module price, which corresponds to the industry revenue, and the cost of installation. The latter coincides with installation added value $AV_{inst}$. According to the investment of a author's relative at the end of 2008 in France and to a quotation for a PV installation at the beginning of 2013, both for a 3 $kW_p$ roof mounted assembly, the installation cost is around $AV_{inst} = 0.70$ € $\cdot W_p^{-1}$ (when necessary euro conversion into dollar is assumed fixed at 1 € = 1.3 \$).

Hence can be calculated the added values *AV* in € for the producer of PV electricity:

$$AV_{PV} = \$_{PV} - \$_{W_p} - 0.70 \cdot F_{W_p},$$

and for the PV industry:

$$AV_{W_p} = \$_{W_p} - \$_{D-W_p}.$$

Each of these quantities must be positive to pay for wages, interests and taxes of each entity.

The added value $AV_{elc}$ for the whole system is examined at the end of section.

Actually the monetary values imposed by markets or government regulations, and usually known, are the unit prices or costs such as the price of one $\text{kWh}_{PV}$, $u\$_{W_p}$, of one $W_p$ of module, $u\$_{W_p}$, and unit cost $u\$_{D-W_p}$ of direct requirements $E_{D-W_p}$.

Hence the added value $AV_{PV}$ is $AV_{PV} = u\$_{PV} \cdot E_{PV} - (u\$_{W_p} - 0.75) \cdot F_{W_p}$, or:

$$AV_{PV} = F_{W_p} \cdot \left( u\$_{PV} \cdot \left( w_{W_p}^{PV} \right)^{-1} - u\$_{W_p} - 0.75 \right). \tag{3.1.1}$$

The value $AV_{PV}$ per $W_p$ depends thus on the unit prices and the inverse of $w_{W_p}^{PV}$.

In the case of the PV industry $AV_{W_p} = u\$_{W_p} \cdot F_{W_p} - u\$_{D-W_p} \cdot E_{D-W_p})$, or:

$$AV_{W_p} = F_{W_p} \cdot \left( u\$_{W_p} - u\$_{D-W_p} \cdot R_D^{W_p} \right). \tag{3.1.2}$$

The added value $AV_{W_p}$ per $W_p$ depends on the unit price of module, unit cost of energy and materials, and the industry rate of consumption $R_D^{W_p}$.

The unit cost $u\$_{D-W_p}$ of requirements such as final electricity for an industry follows the average inflation, at least in Europe (price of about 0.06 € $\cdot\text{kWh}_e^{-1}$ for electricity, generalized at the unit price of other energy forms $E_{D-W_p}$).

### Efficiency Progresses of the Crystalline Si Based PV Industry

**Physics of Photovoltaic Effect** The fundamental element of the PV generator is the photodiode or PV cell. A diode is made of the junction of two differently doped semiconductors. The difference of doping creates a migration of charge carriers (conduction electrons and holes) inducing an intrinsic voltage between semiconductors. It acts like a threshold below which there is no current under an applied voltage. However, when illuminated by a light with photon of sufficient energy, and owing to the intrinsic junction voltage, a photodiode is able to generate a current of charge carriers under a positive voltage, resulting thus in an electrical power.

Various industrial technologies with different semiconductors (crystalline Si and thin films with the junction Tellurium Cadmium or with amorphous Si) exist to make the PV cells. The cells are then connected in series and parallel, and encapsulated to form a module with a surface between 1 to 2 square meters $m^2_{PV}$. Modules are themselves assembled to form larger arrays of PV generators.

**Chain of Operations** In spite of the huge PV industry growth since 2000, the market share of the main and mature technology of cells, the crystalline Si (mono and multi-crystal) processes, has remained similar at around 90% of the overall production in $W_p$, albeit slightly decreasing [23]. The Chinese and Taiwanese companies, mostly responsible for the rise, have preferred the mature and less risky technology.

In this industry the Si metal is first produced from quartz blocks in furnaces similarly to the making of pig iron from iron ore but requiring three times as much energy in the form of electricity and carbon reducers [24].

The metallurgical Si must be then purified by a gaseous fractional distillation (Siemens process in general). The process has been modified thanks to recirculation loops to produce mostly Si and thus to limit the quantity of other much less needed co-products. The purity of the resulting Si can be sufficient for the production of multi-crystal Si modules. It can be also further purified in subsequent operations and cast to form a mono-crystal ingot for the industries of micro-electronic chips and mono-crystalline PV cells.

In the multi-crystal PV case Si obtained from the purification process is melted and cast in molds to form large blocks. The blocks are sawed in squared wafers thanks to slurry based sawing techniques.

Surface treatments, doping processes and layer deposits are then applied to obtain the PV cells. Cells are assembled and encapsulated to form a module.

**Factors of Efficiency of the Industry** A study of the efficiency of the chain of operations with the methodology exposed in the chapter 4 permits to extract the physical and technical factors governing gains on the rate $R^{W_p}$ in the manufacture of modules. Among them are the total mass $\sigma_{Si}$ of PV grade Si required per $m^2_{PV}$ of module, which impacts the consumption per $W_p$ of operations before surface treatment, and the surface power $\eta_{W_p}$ for all operations (another potential parameter, especially in the present time of industry overcapacity, is the factor of utilization, or productive time of the processes relative to their capacity. It is mostly relevant for operations after sawing).

The quantity $\sigma_{Si}$ depends on the losses due to sawing and to the parts of Si too contaminated at the block edges, as well as on the thickness of the Si wafers. Progressively reduced thicknesses of the saw wire and of the wafers, and increased block sizes, have been the main improvements.

The second factor $\eta_{W_p}$ is impaired by impurities and small crystal sizes in the metal (hence the advantage of mono-crystal Si), by light reflections at the module surface and by mismatches between connected cells.

In the early 90s the two factors for the multi-crystal Si modules amounted to $\sigma_{Si} = 2.75\ \mathrm{kg_{Si}{\cdot}m_{PV}^{-2}}$ and $\eta_{W_p} = 118\ \mathrm{W_p{\cdot}m_{PV}^{-2}}$, or a combined value of 23.5 $\mathrm{g_{Si}{\cdot}W_p^{-1}}$ [24]. The cell and saw thicknesses were both about 350 $\mu$m. The Si mass of a cell corresponded to 0.815 $\mathrm{kg.m_{PV}^{-2}}$ (Si density of 2.33 $\mathrm{kg.m^{-3}}$). The whole loss amounted thus to a factor 2.4 of the final quantity. Si block mass was 120 kg.

According to a study of IEA in 2006, a total mass of 15 Mt of purified Si were required in 2005 to manufacture 1 150 $\mathrm{MW_p}$ of cells, or 13.5 $\mathrm{g_{Si}{\cdot}W_p^{-1}}$. The peak power of modules was about $\eta_{W_p} = 125\ \mathrm{W_p{\cdot}m_{PV}^{-2}}$, which leads to a mass of about 1.7 $\mathrm{kg_{Si}{\cdot}m_{PV}^{-2}}$.

Currently the best factories produce 180 $\mu$m thick cells, that is each with a mass of 0.42 $\mathrm{kg.m_{PV}^{-2}}$ of pure Si. Assuming the Si losses due to the distillation processes, the impurities in the block and the cuttings 1.5 time the cell mass (block mass close to 500 kg, saw thickness of about 180 $\mu$m), the requirement $\sigma_{Si}$ amounts to about $\sigma_{Si} = 1.05\ \mathrm{kg_{Si}{\cdot}m_{PV}^{-2}}$. With yield $\eta_{W_p}$ of recent best multi-crystal Si modules at about 150 $\mathrm{W_p{\cdot}m_{PV}^{-2}}$ (average yield is about 140 $\mathrm{W_p{\cdot}m_{PV}^{-2}}$ according to Photon magazine), the total Si consumption of the best factories represents 7.0 $\mathrm{g_{Si}{\cdot}W_p^{-1}}$. Due to the fragility of thin wafers with a size of 15.6 cm during its handling in the downstream operations, the practical thickness limit is close to 180 $\mu$m (by comparison a hair thickness is about 100 $\mu$m thick).

**Efficiency Limits for the Si Crystalline Technology** Besides, gains in the local rates of consumption for the successive operations of PV industry have been probably achieved since 2000, especially the purification operation (its local rate is expressed in $\mathrm{kWh_D}$ per kg of PV grade Si). Values are hard to estimate as the operations and their flows are poorly documented. The gains result mostly from improvements of a mature technology through the economies of scale (bigger reactors, larger blocks and wafers), the heat recovery by process integration - even factory integration - and automation. As a result, without radically new processes, the gains may reach a limit soon.

On the other hand, the improvements were rapidly implemented due to the dramatic rise of the manufacture capacity in the last ten years. Consequently the rate $R_D^{W_p}$ has been probably reduced by a factor of an order of two (due to difficulty to find reliable date no value for the rate is proposed). Gains for labor productivities in worker.hours have been also obtained thanks to the same factors. Both gains lead to a reduction of energy and material costs, and salaries paid by the industry. However, it is questionable that the reductions of the energy and labor rates have followed entirely that of the module price since 2008.

Another limitation of gains for the PV industry is the necessity of the financial amortization of the factories and their equipment and so to operate them over sufficient time. Efficiency becomes thus determined by existing factories as the contribution of the new ones to the production becomes ever more marginal. This situation is occurring right now.

## From 2000 to 2008: Incentives and Large Growth of PV Installation in Europe

Public incentives have been a determinant factor in the evolution of the PV market since 2000, especially in Europe. In April 2000 the Renewable Energy Sources Act (EEG) in Germany made the purchase by utilities of electricity generated from renewable energy to the grid mandatory. The act also instituted a fixed feed-in tariff $u\$_{PV}$ (about $u\$_{PV} = 0.55\,€\cdot\text{kWh}_{\text{PV}}^{-1}$ for roof mounted modules). Utilities were allowed to recoup their costs from their private customers except from industrial customers. Other incentives were tax credits to deduce part of the module purchase.

At the end of 2000 less than 80 $\text{MW}_\text{p}$ of PV panels were connected to the grid in Germany. Four years later the capacity of PV to the grid amounted to 1 100 $\text{MW}_\text{p}$, with about 670 $\text{MW}_\text{p}$ installed during the year 2004 alone. At the end of 2008 the total capacity reached 6000 $\text{MW}_\text{p}$ and is currently more than 30000 $\text{MW}_\text{p}$.

Other European countries followed suit, enacting feed-in tariffs and tax credits for the producers of PV electricity. PV manufacturers could also receive subsidizes in form of low interest loans and grants. All these incentives spurred the PV market. During 2004 a capacity of 1 120 $\text{MW}_\text{p}$ were connected worldwide, largely in Germany and Japan. During 2008 the installation amounted to 6 300 $MW_p$, of which 5 300 $MW_p$ in Europe. Spain was the main market with 2 800 $MW_p$ of modules installed that year, ahead of Germany. During 2007 the Spanish government promulgated a guaranteed price $u\$_{PV}$ like in Germany.

This dramatic growth created a pressure on the PV industry, especially in the production of the solar grade Si. The newspaper Financial Times reported in its edition of 20 November 2006 that the average price rose from 35 in 2004 to 70 \$ per $kg_{Si}$ of PV grade Si in 2006, with transactions that year close to 200 \$ per $kg_{Si}$. With a requirement of 13 $g_{Si}{\cdot}W_p^{-1}$, the cost of Si amounted to 0.92 \$ per $W_p$. The price of benchmark Si crystalline module as reported by the Bloomberg markets magazine in 19 Oct. 2010 rose from 3.2 in 2004 to 4.13 $\${\cdot}W_p^{-1}$ in 2008.

It should be noted that the price of a product for a year, even a well-identified item like multi-crystal Si module when shipped, can differ from one source to another. Thus the IEA 2006 study on the PV industry and market indicated an average price of 4.5 $\${\cdot}W_p^{-1}$ for 2004, which corresponded to a low since the nineties. The newspaper New York Times in its edition of 10 Nov. 2011 reported a whole sale price of PV panels of 3.3 $\${\cdot}W_p^{-1}$ in 2008. Both prices are thus different from those of Bloomberg markets magazine. Looking at these discrepancies, the price can be only a relative indicator issued from the same source.

A French private individual in 2008 bought a PV module at $u\$_{W_p} = 7.5$ €$\cdot W_p^{-1}$ (from the investment of a author's relative). He benefited from a feed-in tariff $u\$_{PV} = 0.57$ €$\cdot kWh_{PV}^{-1}$ agreed by the French national utility *EDF* over module lifetime. If he is located close to Perpignan in Southwest - where solar conditions are similar to Spain ones - his gains over the module lifetime would be $AV_{PV} = 19.5$ €$\cdot W_p^{-1}$ (from eq. 3.1.1). If he is located close to Strasbourg at the German border - solar condition similar to this country - his gains would be $AV_{PV} = 5.5$ €$\cdot W_p^{-1}$.

### From 2008 to 2012: Building the PV Industry Overcapacity

In 2008 the industry had already started to respond to the demand and the consequent rise of module prices. The global annual production of PV cells increased from 1.8 $GW_p$ in 2005 to 7.2 $GW_p$ in 2008. The output from China represented already one third of the total, whereas its production in 2005 amounted to a mere 0.13 $GW_p$. According to the consulting firm GTM research the global manufacture capacity over 2008 was 8.3 $GW_p$ [25].

In 2010 the production of cells amounted to 27.2 $GW_p$, of which 13.0 $GW_p$ came from China, 3.5 $GW_p$ from Taiwan and 2.7 $GW_p$ from Germany. The industries of all countries experienced output growth, but not at the same intensity [23]. The global capacity was close to 30 $GW_p$ [25]. In the demand side a capacity of 16.8 $GW_p$ were installed during 2010, a rise of more than 100% from 2009, of which 13.4 $GW_p$ took place in Europe.

In 2011 the manufacture production and capacity were about 35 and 60 $GW_p$ respectively, while worldwide module installation corresponded to nearly 30 $GW_p$. The European market represented 22 $GW_p$. Over 2011 the Chinese industry had an estimated capacity of 39 $GW_p$. Major manufacturers were Chinese like the company Suntech with an output of 2.2 $GW_p$. Nevertheless, the German group Q-cells reported its highest output at 0.80 $GW_p$.

Estimates of the manufacturing capacity for 2012 put the number to 72.5 $GW_p$, of which about 50 $GW_p$ were located in China [25]. The service society NPD Solarbuzz in its March 2013 report on the solar market indicates a global module installation for 2012 of just 29 $GW_p$, of which 16.5 $GW_p$ took place in Europe.

Two opposing trends has occurred since 2008:

- reduction of the financial incentives in Europe,
- the fall of module price due first to efficiency gains in manufacturing and then harsh competition.

The change of public policy in relation to PV generation is best illustrated with the developments occurred in Spain. After the promulgation of guaranteed feed-in tariffs, and faced with the growing deficit of utilities - about 5 G€ in 2008 (Reuters Sept. 23, 2008) resulting in large part from the PV policy -, the Spanish government had to backtrack. Moreover, the country was hardly hit by the financial crisis at the end of 2008, especially in its banking and construction sectors. The feed-in tariff was slashed by one fourth and, above all, subsidizes were capped to a yearly installation of 0.50 $GW_p$. As a result the capacity installation fell to 0.02 $GW_p$ in 2009 from 2.8 $GW_p$ in 2008. However, the modifications were insufficient as the deficit of utilities reached 30 G€ at the end of 2010. Besides, the pending obligations to the producers of PV electricity have accumulated to 125 G€. The government decided at the end of 2010 to apply a retroactive cut to the previous "guaranteed" agreements. In January 2012 it declared a moratorium in installation.

This reversal of policy and the consequences repeated themselves in the Czech republic with the installation of 1.50 $GW_p$ in 2010 and only 0.006 $GW_p$ in 2011 [26]. Italy experienced somewhat with this change (from 9.3 $GW_p$ in 2011 to 3.2 $GW_p$ in 2012). Greece (with about 0.85 $GW_p$ in 2012) is probably next in the list after its government reduced the feed-in tariffs. Even Germany intends to limit the yearly installation at less than 4 $GW_p$, instead of the recent trend of 7.5 $GW_p$ (feed-in tariffs for a 3 $kW_p$ assembly at 0.16 €$\cdot kWh_{PV}^{-1}$ in April 2013 against 0.25 €$\cdot kWh_{PV}^{-1}$ for 2012).

Even reduced, but still in place, the tariffs were interesting due to the fall of module prices $u\$_{W_p}$. According to Bloomberg news agency from its benchmark of Si crystalline modules, the price decreased to $u\$_{W_p} = 1.9\ \$\cdot W_p^{-1}$ in 2010 and less than 1 $\$\cdot W_p^{-1}$ in 2011 and after. Assuming a module price at $u\$_{W_p} =$ 1.0 €$\cdot W_p^{-1}$ for the final purchaser of a 3 $kW_p$ assembly, and a feed-in tariff at $u\$_{PV} = 0.25$ €$\cdot kWh_{PV}^{-1}$, the gains in Germany over the module lifetime would be $AV_{PV} = 4.3$ €$\cdot W_p^{-1}$ (from eq. 3.1.1). The total cost includes that of the installation, which has experienced little or no reduction. At a tariff $u\$_{PV} =$ 0.16 €$\cdot kWh_{PV}^{-1}$, the gain falls to $AV_{PV} = 2.1$ €$\cdot W_p^{-1}$. With same unit prices an investor in Spain would still have $AV_{PV} = 6.0$ €$\cdot W_p^{-1}$ due to better solar irradiance.

After 2010 the discount on the module price was obtained at the expense of PV industry margins. Nearly all Western countries manufacturers were losing money unable to pay their debts from equipment investments [25]. Many went bankrupt, like Q-cells in April 2012. The major public petroleum company British Petroleum closed its subsidiary BP Solar in spite of thirty years of activity. According to the market research consultant IHS iSuppli a contraction of the number of companies in the upstream Si industry - from raw solar grade Si to encapsulated modules - is expected to reduce this number from 750 in 2010 to 130 at the end of 2013 (reported by M. Hall in PVmagazine news, 11 Jan. 2013).

The demise of the western PV industry triggered a trade war with China, accused of economic dumping. Indeed its industries have benefited from low interest loans and grants due to government policy, without mentioning the smaller costs of labor and Si feedstock. The Chinese industry was tailored towards the export market. In May 2012 the US state imposed import tariffs on Chinese modules higher than 34% of its price. The Europe Union is preparing a similar measure.

China has started to develop an internal market since 2011 in view of mounting difficulties in the export market (2.2 $GW_p$ in 2011, 4.6 $GW_p$ in 2012). Its own industry is also suffering from the overcapacity and the fall of prices. Chinese factories owe 18 billions \$ of loans to state owned banks and are losing up to 1\$ for every 3 \$ of sales in 2012 [25]. They started to close part of their production lines. The first global manufacturer in 2011, Suntech, went bankrupt in March 2013.

### Summary and Concluding Remarks about Electricity Production from PV

**Current Situation and Prospects** Public subsidies triggered a demand of PV generators and hence an industry development. Real gains on efficiency were obtained at the level of processes of PV manufacture. They explain partially the decrease of the price of PV module after 2008. The rest results from the industry overcapacity leading to harsh competition and sales at losses.

In spite of this dramatic fall of prices, when the subsidies are removed investors lose their interest in PV modules. A whole sale price of electricity at $u\$_{PV} = 0.05$ €$\cdot \mathrm{kWh}_{\mathrm{PV}}^{-1}$ would imply for a German producer a negative added value ($AV_{PV} = -0.55$ €$\cdot \mathrm{W}_{\mathrm{p}}^{-1}$; whole sale price may be even negative at noon when PV output in a country is maximal), barely positive for a Spanish producer.

The risk exists that the subsidies and the consequent industry demise have blocked the development towards a true cheap PV captor. The incentives have favored a proven technology - manufacture of crystalline Si modules - and accelerated its improvements towards its physical and technical limitations. But the Si module is still not sufficiently cheap to compete with conventional generators without some public helps (except in regions with a lack of transmission lines and/or an expensive electricity). A cheap sun captor requires a radically new technology. However, the current financial situation of industry and government makes it difficult for investments to develop this new technology.

**How Physics and Engineering Intervene within the System Added Value** solar radiation is ubiquitous on the Earth's surface and almost permanent at the scale of human history. However, it is a variable and diluted form of energy producing the same characteristics on PV output.

The losses due to the electricity regulation to smooth variation is taken into account by the factor $w_{PV}^{elc}$ (the larger the losses are, the higher $w_{PV}^{elc}$ is).

Dilution is measured by the total output of PV modules, $\eta_{W_p}/w_{W_p}^{PV}$, in $kWh_{PV}$ per $m_{PV}^2$ of surface (the higher the dilution is, the lower the factor is). A PV captor converts thus radiation at an average power of between 15 and 35 $W_{PV} \cdot m_{PV}^{-2}$ over a lifetime of about 25 years, depending on the solar irradiance in its location. We can compare this density of energy to that of a recent onshore wind turbine. In terms of volumetric energy density of the resource, wind displays the lowest one among the conventional forms of primary energy (see the paragraph 3.3.2 in the section 3.3). The 2 $MW_p$ Vestas turbine V80 possesses three 40 m long blades, or presents to wind a 5 000 $m^2$ surface of capture. Assuming a load factor of 25% for this turbine (Danish onshore average), its power averages 100 $W_e$ per $m^2$ of its blade surface over a lifetime of about 25 years too, that is a higher density than that of PV captor.

As a result the captor per unit of its surface or peak power must requires very low quantity of energy and materials (or small rate $R_D^{W_p}$), as well as small amounts of economic resources such as worker.hours and capital investments from high added value units.

Let's traduce these requisites by two equations.

The added value $AV_{elc}$ for the whole system *elc* results from the balance between its revenues and costs, $AV_{elc} = \$_{elc} - \$_{D-W_p} - \$_{D-eqp}$. By introducing unit prices and energy flows we have:

$$AV_{elc} = u\$_{elc} \cdot E_{elc} - u\$_{D-W_p} \cdot E_{D-W_p} - u\$_{D-eqp} \cdot E_{D-eqp}.$$

By introducing the industry consumption rates to manufacture equipment, and factors of conversion:

$$AV_{elc} = E_{elc} \cdot \left( u\$_{elc} - u\$_{D-W_p} \cdot R_D^{W_p} \cdot w_{W_p}^{PV} \cdot w_{PV}^{elc} - u\$_{D-eqp} \cdot R_D^{eqp} \cdot w_{PV}^{elc} \right). \tag{3.1.3}$$

On the other hand, by taking into account eq. 3.1.1 and 3.1.2, and including similar contributions from subsystem *tr*, the added value $AV_{elc}$ also corresponds to the sum of the added values of its subsystems,

$$AV_{elc} = AV_{PV} + AV_{W_p} + AV_{tr} + AV_{eqp} + AV_{inst}. \tag{3.1.4}$$

Each of them must be positive to pay for their wages, interests and taxes, in absence of subsidize.

The eq. 3.1.3 expresses the origin of the added value while eq. 3.1.4 how it is employed. Better labor and capital productivities reduce the added value to obtain.

In the case of PV generator thanks to mandatory purchase and advantageous tariffs electricity production has become artificially a high added value unit. However, electricity at the user outlet - residential or industry - should remain cheap (low prices $u\$_{elc}$, typically 0.06 € $\cdot \text{kWh}_\text{e}^{-1}$ for industry). Consequently the whole electricity sector is still a low added value unit. Subsidizes are necessary to offset the high value $AV_{PV}$, but they have to be temporary to avoid a drain of money from utilities and/or government treasuries.

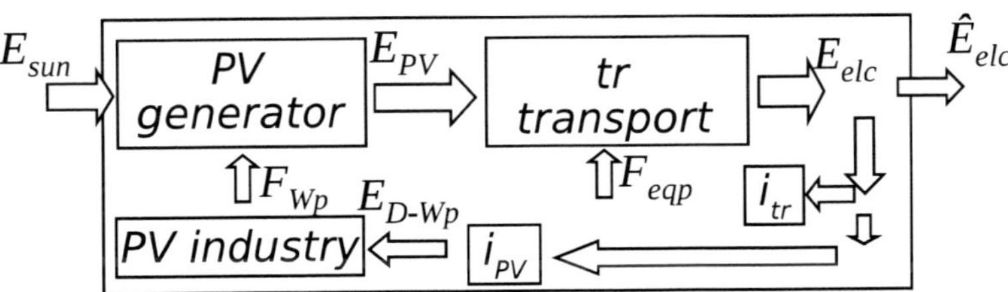


Figure 3.1.6. Diagram of flows of electricity production from solar input $E_{sun}$ to output at user level $\hat{E}_{elc}$, adapted from Fig 3.1.5 in case of a self reliant system. The system provides from its gross output $E_{elc}$ its own consumptions such as the flow $E_{D-W_p}$ to manufacture $F_{W_p}$ modules. Type of requirements and constraints on their delivery impose intermediate operations such as $i_{PV}$.

**Self-Reliant Energy System** Either by omitting high added value equipment like machine tools or by including their industries in the system *elc*, we assume that direct requirements $E_D$ consist only on basic products, electricity, processed fuels and chemicals. All their unit costs $u\$_D$ expressed per unit of energy are also supposed similar and equivalent to the unit price $u\$_{elc}$, $u\$_{D-eqp} \equiv u\$_{D-W_p} \equiv u\$_{elc}$. The eq. 3.1.3 becomes thus:

$$AV_{elc} = E_{elc} \cdot u\$_{elc} \cdot \left(1 - R_D^{W_p} \cdot w_{W_p}^{PV} \cdot w_{PV}^{elc} - R_D^{eqp} \cdot w_{PV}^{elc}\right) = \hat{E}_{elc} \cdot u\$_{elc}.$$

A positive added value without subsidize or deficit depends on the physical surplus fixed by the expression in parentheses of above equation.

Actually the case described by the equation is equivalent to the energy system *elc* modified to provide with its gross output $E_{elc}$ its own direct requirements $E_{D-j}$. It corresponds to the self-reliant system as described by Fig. 3.1.6. Intermediate operations $i_{PV}$ and $i_{tr}$ are introduced to adapt the gross flow to constraints on industry requirements.

The economic viability of the system of Fig. 3.1.6 is first and foremost dependent on its energy existence, i.e. a positive $\hat{E}_{elc}$.

### 3.1.4. Energy Efficiency as a Predictive Tool for Economic Growth?

**Drawback of the Energy Intensity as an Indicator of Energy Efficiency and of Future Growth** In view of some lack of correlation between the rate $R_D$ and the energy intensity $I_D$ in recent data of the American economy, is it still appropriate to use the national accountability and the price of a product to deduce its energy consumptions?

Thanks to the economic input and output table for each economic unit - along with its consumption of electricity - the national accountability offers a means to convert a price to an amount of energy. The price is related to the revenue of the economic unit used to manufacture the product. From the input and output table you can deduce in principle the annual energy consumption of the whole unit and subsequently the consumption per unit of its revenue. Hence it can be determined what the energy per unit of product consumed by the sector is.

Input and output tables represent the sums of sales and costs over a year and consequently are based on averages. However, due to the sometimes rapid price changes (an extreme case is the crude oil price during 2008, which changed by nearly a factor 5. Market of PV modules offers another example), and because of the differences of price paid by the various customers according to their importance, a unique price over a year does not exist. It appears difficult to connect prices to the more stable energy consumption per product. The price is more related to the market situation of the product in the short and medium terms, at least for the high added value sectors. Moreover, very often the economic unit offers other goods or services with different prices. Among the low added value sectors, the refineries produce various petroleum products from liquefied petroleum gas to gasoline to lube oil of which prices also depend on demand. A high unit price for a product will artificially increase its energy consumption when using the average coefficient of conversion from the input and output table.

Furthermore a complete energy balance necessitates the exhaustive list of materials required by the industry and the energy consumptions per unit of each material. The latter is deduced from the monetary and material flows of the industries producing the material, like for the industry under study. Consequently the same approximate method of correspondence between monetary values and energy consumptions must be employed. Moreover the physical lists of materials and fuels consumed provided by economic surveys for each economic unit can contain some important errors and/or oversights as observed with the US economic census for the petroleum refinery and the N fertilizer industry. It is also only provided every five year.

Interestingly, if the material data are erroneous the corresponding monetary values can be correct. The purpose of the national accountably is to establish the monetary flows between the different units of the economy, and to a large degree neglects the material flows. When taking into account all the chain of upstream sectors, energy appears nowhere in the overall financial balance. The costs to extract and process it are only expressed in terms of wages, taxes and capitals. From the economic point of view, whatever the type of primary energy that nature supplies is deemed to be free.

In brief the deduction of a rate $R_D$ of consumption for an industry from the price of its product provides at best an indication of order of magnitude.

**Use of the Energy Rate as Indicator** What about the opposite, *i.e.* the rate $R_D$ giving insights on the economic viability of an industry expressed by its added value?

Obviously some short-term factors would be absent such as the development of salaries, demand, taxes, subsidies, and macro political events affecting the production.

But if we consider the longer term, which may be more important for an industry with large investments and long times of return of investment, the rate $R_D$ and the variables influencing it can be useful. Large improvements of current processes and above all new processes for the same operation take a long time to implement at a large industrial scale, even once a first operational pilot plant has been built. The energy balance of an industrial pilot of the system can be already ascertained, whereas the financial balance is harder to establish. The selected processes should fulfill the economic constraints mentioned in the chapter 2. On the other hand, a system which requires too many materials and dissipates too much energy along a lot of complex processes due to fundamental limits, like likely $H_2$ as a means of electricity storage for a local grid [27], are not viable. The gas presents a too low volume content of energy in normal conditions to be competitive.

A technical study is therefore more able to make accurate predictions. However, it requires a very detailed study of the operations of a known or promising new industry based on its processes, whether already established or at the industrial pilot scale. This information may be difficult to obtain for lack of industry participation or for fear to share proprietary data. It is not possible to only rely on producer statements. Moreover the data must be checked against established knowledge.

### 3.1.5. Summary and Conclusions about Economic Relevance

#### Origin and Use of Added Value. Link with Efficiency

Analyses achieved in this section about interplay between economy and energy efficiency can be summarized by a set of two equations.

For any system or economic unit $M$, given its output $L_{out}$, the unit price $u\$_{out}$ of its products, the unit cost $u\$_D$ of its direct requirements $E_D$ of energy and materials, and the rate $R_D^M$ of consumption of $E_D$ by the system, its net income $AV_M$ are provided by:

$$AV_M = L_{out} \cdot \left(u\$_{out} - u\$_D \cdot R_D^M\right). \qquad (3.1.5)$$

The system ranges from an industry like petroleum refinery to the economic activity of a country. The added value of the latter corresponds to its gross domestic product GDP. To account for the diversity of requirements $E_D$ - from final energies $EF$ to equipment and to services like assurance -, the rate $R_D^M$ is in form of matrix while the cost $u\$_D$ is a vector. Energy requirements to produce services or materials can be taken into account by enlarging the system to include the industries or sectors producing such items.

On the other hand the added value $AV_M$ for$M$ from eq. 3.1.3 is distributed as:

$$AV_M = \text{salary} + \text{interests} - \text{deficits} + \text{taxes} - \text{subsidizes}. \qquad (3.1.6)$$

#### Evolution of Added Value and Efficiency for Different Sectors

Economy of a country before the middle of twentieth century was dominated by the values generated by industries of basic products such as energy, food, chemicals, building materials... In the last one or two centuries they made large progresses to reduce their rate $R_D^M$, both in energy and materials, and to increase their output (eq. 3.1.5). In the same time their labor productivity rose, resulting in lower salary cost per unit output (eq. 3.1.6).

The section 3.1.1 shows the case of the ammonia industry, which, over the twentieth century, reduced its rate for gas by a factor 5 and that for steel within equipment by at least six. During the same period it increased its output by a factor 150. Unit price $u\$_{HHV}$ of gas consumed by the industry also diminished owing to similar efforts in the upstream industries of gas processing and transport. All these gains permitted for these basic industries to decrease considerably their unit price $u\$_{out}$.

Their weight in the GDP of the economy of a country such as the USA fell due to their gains and the emergence of large added value sectors. These new ones range from manufactures of consumer goods to services related to culture and leisure. They thrived on the lower costs of final energy like electricity and other basic products. Their main costs became purchases from other high added value industries. As a result their added value and their energy intensity $I_D$, final energy required per unit of added value, were increasingly unrelated to their rate $R_D^M$ of consumption of basic products.

But even some energy intensive industries have seen their profits mostly dominated by prices of their outputs and inputs.

Over the period between 2002 and 2007, prices of some energy such as crude oil and gas experimented a twofold - or larger - rise. The rise represented, however, few percent points of GDP of a developed country like the USA or of the added value of the new economic units. Over the period all accepted to pay the incremental cost, making some little arbitrage with a still cheap form of energy like electricity, in order to maintain their output $L_{out}$ and subsequently their revenue $\$_{out}$. In spite of the low monetary importance of basic products they are physically indispensable, which is only captured by the rate $R_D^M$. Meanwhile, net income $AV_M$ of some basic industries such as petroleum refinery and ammonia industry has been governed by the large and short-term variations of energy prices, while their output $L_{out}$ and rate $R_D^M$ have remained relatively constant. The low present influence of rate $R_D^M$ for these industries is reinforced by the small gains obtained in recent factories due to fundamental limitations, as observed in the case of the ammonia production.

The study of photovoltaic based electricity production has offered a complex and contrasting view on the added value of a system. The whole system *elc* comprises subsystems from the manufacture of PV module to electricity production and to electricity delivery at the end user. Because price $u\$_{elc}$ of delivered electricity is still low, the energy system is part of basic industries with a low added value.

However, some of the system actors - electricity producers and manufacturers of modules - have been rewarded with large profits, at least until 2008 for the latter ones, thanks to the high unit prices of their products, PV electricity and modules. PV industry also benefited from large output - one thousand fold increase in ten years -, and by the reduction of its rate $R_D^{W_p}$ (order of two probably although data are sparse on processes). Low added value at the level of the whole system was made possible only by subsidies (price fixing) or increase of electricity price for some consumers. Due to PV industry overcapacity and harsh competition after 2008, module price has crashed leading to a deterioration of PV industry added value according to eq. 3.1.5, resulting in deficits, unpaid loans and bankruptcy to be compatible with eq. 3.1.6.

The system *elc* is an energy system consuming electricity to produce electricity. Like for other primary energy industries its feedstock - solar radiation - is free. If we assimilate all its consumption to electricity (by including the industries providing items such as machine tools and chemicals, and by imaging processes operating only with electricity) unit price and cost are similar ($u\$_{elc} \equiv u\$_D$). This approximation leads to a simpler added value from eq. 3.1.5:

$$AV_{elc} = L_{elc} \cdot u\$_{elc} \cdot \left(1 - R_D^{elc}\right).$$

It is actually the added value of a self-reliant system (see the chapter 5). Before looking at economic relevance physical existence must be assured with rate $R_D^{elc}$ lower than one. Economic profitability requires further reduction of the rate. Economy on the long term is thus dependent on fundamental limits and technical constraints.

In turn choice of processes must assure gains in other parameters of the added value: large scale production and sale $L_{elc}$, which require reliable processes and good quality products, high labor productivity, low amount of investment to reduce interest to pay, complying with regulations and labor laws to avoid excessive taxes... (see the section 2.3.1 of the chapter 2). Industry would not thus rely on continuous deficits and/or subsidizes (eq. 3.1.4). In these conditions unit price can be lowered to make the product, electricity here, attractive and to benefit the rest of the economy.

## Case of a Profitable but Inefficient Energy Industry

To conclude this section about economy we show a situation where an energy system requires more energy than it produces (in the configuration of self-reliant system), whereas it presents a positive financial balance owing to the use of external cheap energy or material, and an advantageous price for its product. The situation is similar to that of the PV module manufacturer and electricity producer, except that, in absence of reliable data on the efficiency of module manufacture, we have not decided on its energy balance.

The example represents an extreme case of what is encountered in some oil producing wells in California in the USA (Oil and Gas Journal, 12 Dec. 2005, *oil producers slow to apply energy efficiency*). Oil reservoir is at $h = 2000$ m depth and requires water injection to maintain pressure and produce some oil. Well output is made of 99% water and 1% oil in weight. Lifting one tonne of fluid requires an electricity consumption of (2 to 5)·9 800·$h$ $J_e \cdot t^{-1}$, according to field data. In our case the consumption amounts to between 40 and 100 $MJ_e$ for one tonne of fluid, of which energy content owing to oil (10 kg oil) is about 420 $MJ_{PCI}$. The rate $R_D^{well}$ of direct consumption is 0.093 to 0.23 $J_e \cdot J_{PCI}^{-1}$, or 0.54 to 1.34 $GJ_e \cdot b^{-1}$ (one barrel of typical crude has a heat value of about 5.8 $GJ_{PCI} \cdot b^{-1}$).

Assuming a self-reliant energy system - due to the absence of nearby grid - by operating an oil fueled generator with yield of 20%, the dissipation rate is $R_{diss}^{well} = 0.46$ to 1.15 $J_{PCI} \cdot J_{PCI}^{-1}$. In the worst case the system cannot exist ($R_{diss}^{well} > 1$), like the deep coal mine with Newcomen steam engines (section 5.1.1 of the chapter 5).

Now, the well pumps are connected to a grid, which supplies electricity at a unit cost $u\$_D = 0.08$ $\$ \cdot kWh_e^{-1}$ or 22 $\$ \cdot GJ_e^{-1}$. On the other hand, well owner receives for the oil a unit price at $u\$_{oil} = 100$ $\$ \cdot b^{-1}$ or about 17 $\$ \cdot GJ_{PCI}^{-1}$.

The well added value is (equipment cost is omitted):

$$AV_{well} = u\$_{oil} \cdot E_{oil} - u\$_e \cdot E_e \quad \text{or,} \quad AV_{well} = E_{oil} \cdot \left(u\$_{oil} - u\$_e \cdot R_D^{well}\right)$$

The added value $AV_{well}$ is positive and ranges from 12 to 13.1 $\$ \cdot GJ^{-1}$, or 70 to 88 to $\$ \cdot b^{-1}$.

An efficient energy system pays for an inefficient one, which only a process based study can reveal.

## 3.2. Population, Consumer Behavior and Energy Saving

Progresses in the energy efficiency of processes reduce the energy requirements per unit of a product and so the overall energy consumption, provided production remains constant. However, viewed at the scale of a country or the world the total consumption of final energy or products rise with population growth and standard of living.

Consumption will be equally influenced by the behavior of the average consumer and his willingness to conserve energy and/or to make products last longer.

### 3.2.1. Population Growth and Increase in Energy Consumption

This section deduces from official statistics for some years between 1929 and 2009 the correlation of population growth and energy production/consumption. It studies the development of per capita consumption for the world and for selected countries or groups of countries. The human basic requirements of energy is also assessed as a comparison. Lastly it intends to determine the mechanisms between growths of both population and energy production, or which one is the determinant of the other.

**The Correlation between Population and Energy Consumptions since 1929**

**Table 3.2.2. Populations and primary energy and electricity consumptions in the world, the USA and France for 1929, 1950, 1973 and 2009**

| | World | | | USA | | | France | | |
|---|---|---|---|---|---|---|---|---|---|
| | pop. Gcap | prim. $Gtoe_{EP}$ | elc $Gtoe_e$ | pop. Mcap | prim. $Mtoe_{EP}$ | elc $Mtoe_e$ | pop. Mcap | prim. $Mtoe_{EP}$ | elc $Mtoe_e$ |
| 1929 | 2.0 | 1.54 | 0.022 | 122 | 620 | 7.9 | 41 | 72 | 1.2 |
| 1950 | 2.5 | 2.30 | 0.075 | 150 | 825 | 28 | 41 | 58 | 2.9 |
| 1973 | 4.0 | 6.10 | 0.44 | 212 | 1 750 | 147 | 52 | 180 | 13 |
| 2009 | 6.8 | 12.2 | 1.44 | 308 | 2 160 | 320 | 65 | 256 | 42 |

**Presentation of Data** Table 3.2.2 shows the evolution of the global supply of primary energy and of final electricity for the world and two developed countries - the USA and France -, for four recent years, 1929, 1950, 1973 and 2009.

Because of its disproportional share of world energy consumption in relation to its share of world population (40% of the global consumption in 1929 by just 6% of the population) and also of energy production - although slightly less so in recent times - the USA distorts the energy statistics at the world level. To study the changes of global per capita consumption the USA contributions are removed.

France represents an intermediate country between the USA and the rest of the world, although it is part of developed countries. It can be used as a reference point and aim for the rest of world in absence of severe constraints.

For the years up to 1950 all the data are provided by a report of the statistical office of the United Nations published in 1952 [28]. For 1973 and 2009 data have been compiled by the International Energy Agency IEA, an agency of the Organization for Economic Cooperation and Development [29]. Fig. 3.2.1 and 3.2.2 show the global supplies of primary and final energies in 1973 and 2009 according to this source.

Information is also available at the national agencies in charge of energy statistics such as the Energy Information Administration, which compiles the Annual Energy Review for the USA [16], and *l'Observatoire de l'énergie* in France [30]. Information about the population is also provided by these departments, or otherwise retrieved from US census bureau.

The data have been corrected for the convention differences between the various sources[2]. Due to the difference of conventions, the complexity of the energy systems, the sheer number of producers and consumers, and errors of conversion, the values reported in Table 3.2.2 and in Fig. 3.2.1 or Fig. 3.2.2 have at best an accuracy within a few percent points.

The primary energy represents the heat value of all forms of energy extracted or produced by humans in mines, fields, forests and power plants. It also includes an approximate quantity of the biomass used directly by households especially in poor countries (see below).

Final electricity represents the electricity at the end user level, reduced by the producers own requirements and the loss of transport and distribution (the overall two contributions are estimated at 16% of the gross electricity genera-

[2]See the appendix for details on these various conventions and the factors of conversion

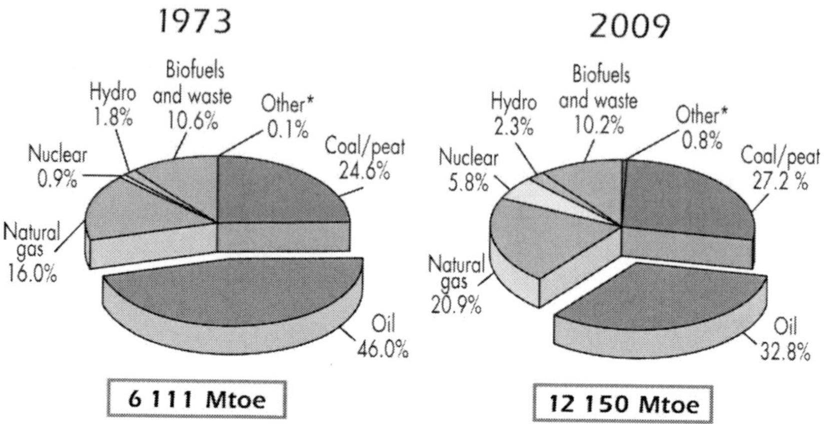


Figure 3.2.1. World supplies of primary energy in 1973 and 2009 according to the International Energy Agency [29] (toe or ton oil equivalent = 41.86 GJ).*Other includes geothermal, solar, wind, heat, etc.

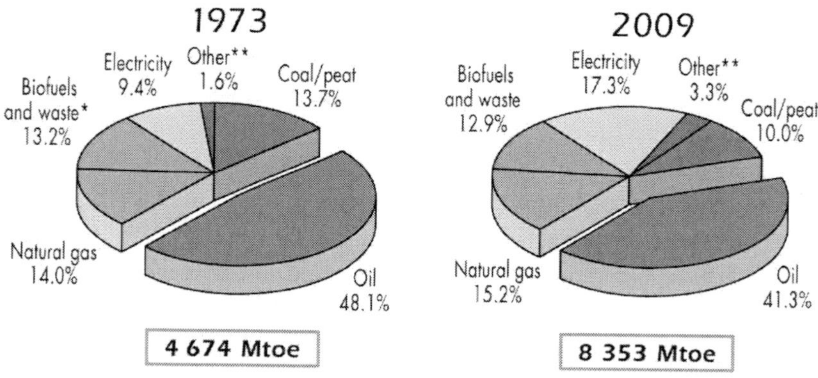


Figure 3.2.2. World consumptions of final energy in 1973 and 2009 according to the International Energy Agency [29].*Data prior to 1994 for biofuels and waste final consumption have been estimated.**Other includes geothermal, solar, wind, heat, etc.

tion, see the appendix). It is part of the final energy as shown in Fig. 3.2.2. Electricity use represents an interesting indicator because of electricities growing importance in all sectors of the economy - as shown in Fig. 3.1.2 of the par 3.1.2 -, as well as individual households. This importance also stems from its ability to be produced from all primary energies and to substitute for all other final energies. Electricity drawback is the lack of large scale storage technologies, although its production can be regulated to match the demand at a country scale.

The present analysis is intended to determine a trend over periods between selected years. Because each set of data corresponds to a particular year, as provided by the sources, it may be distorted by major events during the year or just before. In fact the years 1929 and 1973 are situated just before major economic crises and so can present a local peak of consumptions. The year 1950 falls within the postwar years (5 years after the end of the second world war). The year 2009 experienced the fallout of a major economic crisis, which is assumed to have decreased consumptions. The decrease was actually most pronounced in developed countries and represented a few percent points (see for instance Fig. 3.2.3(b)).

**Analysis of Data** From Table 3.2.2 we observe that all variables - populations and consumptions - have experienced an increase in average, in spite of the economic crises and wars during the 80 years covered. The data show the impact of the second world war - and maybe the crisis of 30s - only in the data relative to the French population and its primary energy consumption (but not to its electricity one). Growth has probably commenced before 1929, but the periods covered show the highest rates per year as averages.

As far as the growth of the world population is concerned, its rate was about 1.0%$\cdot$y$^{-1}$ between 1929 and 1950, 2%$\cdot$y$^{-1}$ between 1950 and 1973, and 1.5%$\cdot$y$^{-1}$ thereafter. The growth for the two developed countries was generally lower, albeit still important for the USA thanks in part to immigration (1.0%$\cdot$y$^{-1}$ after 1973). By comparison, from estimates given by the US census bureau, the world population increased from about 0.79 G capita in 1750 to 1.65 G capita in 1900, or a rate as low as 0.5%$\cdot$y$^{-1}$. From the year 1000 to 1750 the average rate was a mere 0.1%$\cdot$y$^{-1}$.

The growths in energy consumptions are even more dramatic: at the world level the rates for primary energy were successively 2.0, 6.0 and close to 2.0 %$\cdot$y$^{-1}$ for the three periods under study. The trend was similar for the two developed countries, albeit lower (with the exception of France between 1929 and 1950 ; the consumption caught up after 1950). Before 1929 these countries had already a large base of consumption.

Uses of electricity experienced and still experience the largest rates: succes-

**Table 3.2.3. Per capita primary energy and electricity consumptions in toe$\cdot$ cap$^{-1}$ in the world, the USA and France for 1929, 1950, 1973 and 2009**

| | World | | USA | | World w/o the USA | | France | |
|---|---|---|---|---|---|---|---|---|
| years | prim. | elc. | prim | elc. | prim. | elc. | prim. | elc. |
| 1929 | 0.77 | 0.011 | 5.0 | 0.065 | 0.52 | 0.007 | 1.7 | 0.03 |
| 1950 | 0.91 | 0.030 | 5.6 | 0.19 | 0.61 | 0.020 | 1.4 | 0.07 |
| 1973 | 1.55 | 0.11 | 8.3 | 0.70 | 1.15 | 0.077 | 3.5 | 0.25 |
| 2009 | 1.80 | 0.21 | 7.0 | 1.04 | 1.54 | 0.173 | 4.0 | 0.64 |

sively 6.0, 8.0 and 3.4%$\cdot$y$^{-1}$ for the three periods at the world level. Large rates are also observed in the developed countries. Between 1950 and 1973 the rates for the USA and France were thus 7.5 and 7.0%$\cdot$y$^{-1}$ respectively.

The period between 1950 and 1973 stands out in this growth analysis. Known as *the Golden fifties and sixties* it coincided with the largest economic expansion in human history, especially for developed countries. Importantly the households participated to this consumption, which served to satisfy not only their needs but also their wants for comfort. They acquired modern durable consumer goods such as cars, washing machines, fridges, radio or TV set... to ease everyday tasks and to provide access to new means of leisure and/or culture.

Agricultural yields have also increased dramatically. The corn yield in the USA rose from about 1.5 to more than 5 tonnes per hectare. The progress in industrial processes such as ammonia production enabled the construction of large capacity facilities and their productivity increase (see section 3.1.1 for ammonia). The discovery and the start of production of the largest ever oil fields occurred just before or after 1950. Oil production represented less than 540 Mtoe$_{EP}$ in 1950, while it amounted to 2 800 Mtoe$_{EP}$ in 1973, corresponding to a growth rate of nearly 7.5%$\cdot$y$^{-1}$. All of these industrial developments played a critical role in the observed growths.

As a result of the differential of growth between population and consumption, Table 3.2.3 shows an increase of the per capita consumption over time. The population growth has been accompanied by higher production and consumption but the trend is divergent, especially when starting from a low level of consumption. Only two cases depart from this trend: primary energy consumptions in France before 1950, due to the second world war, and in the USA after 1973. In the latter case the oil crises after 1973 pushed to energy savings at large scale (see the next section) and substitution for oil use like in the electricity sector. The strong increase of per capita electricity use is observed in all places, the world outside the USA and the developed countries, and for each period, with the special features of the the fifties and sixties.

Interestingly, during the period after 1973 the per capita annual consumption of oil in the world has diminished from 0.72 to 0.59 $toe_{LHV}{\cdot}(cap{\cdot}y)^{-1}$ (Fig. 3.2.1). But this per capita decrease was largely offset by the contribution of other primary energies, coal, gas and nuclear. Moreover, the per capita usage of electricity, which is of more utility to the users than oil at same heat value, went from 0.11 in 1973 to 0.21 $toe_e{\cdot}cap^{-1}$ in 2009.

Before looking at the mechanisms to explain these correlations, let's determine the basic requirements of humans in case access to energy is limited. It provides a baseline to compare with values shown in Table 3.2.3. This basic contribution is included in the consumptions of primary energy given in Table 3.2.2.

**Basic Requirements - Non Commercial Energy**

The report of statistical series published by the United Nations in 1952 suggested a method to estimate the lowest per capita requirements. At that time the population of lot of countries could not afford what the report calls commercial energies, fossil fuels and hydro-electricity. As in the past they had to rely on local fuels, fuel wood, agriculture wastes like bagasse of cane, lumber mill wastes, dung, or even peat like in Ireland and some parts of the URSS. These forms of energy were used by humans for their household purposes (heating, cooking...), and were rarely reported.

The UN report found it necessary to indicate at least a minimum for consistency between countries, some of which declared the consumptions of these non commercial energies. It used the data given by these few countries for the year 1949, as well as a study of the energy purchased by some Indian households wealthy enough to access some commercially available energy, however, limited for their basic needs. India benefits from a benign weather, and at the same time, available wood lots or other biomass are deficient. Consequently heating requirements are minimal. Hence these energy consumption figures exemplify the notion of minimal requirements. Including the addition of some wastes, the basic need varied from 0.15 to 0.25 $toe_{EP}$·.(cap·y)$^{-1}$ with an average of 0.17 $toe_{EP}$·(cap·y)$^{-1}$. A special category was added to the statistics in order for the population of each country to fulfill at least this minimum per capita consumption. Data from the few countries which relied largely on this type of energy and with a somewhat equivalent benign climate show energy use of the same magnitude (the case of Greece, Portugal, Hungary and El Salvador in 1949). On the other hand Finland reported for its consumptions of fuel woods and lumber mill wastes a much higher rate, 1.5 $toe_{EP}$·(cap·y)$^{-1}$. Heating requirements are very high for this sub-polar country and abundant forests can provide them. A similar observation can be made for Ireland with its use of peat, which amounted to 0.71 $toe_{EP}$·(cap·y)$^{-1}$.

In fact the whole estimate of non commercial energy for 1949, the only year in which it was done, gave a total of 415 $Mtoe_{EP}$. Its breakdown was 30 for peat and bagasse, 95 for fuel wood, 25 for wood wastes and 265 $Mtoe_{EP}$ of unreported fuels estimated according to above model. The total is thus equivalent to a per capita use of 0.17 $toe_{EP}$·(cap·y)$^{-1}$. The primary energy consumption given for the years 1929 and 1950 in Table 3.2.2 include these non commercial contributions. It is noteworthy that the amounts of biofuels and wastes reported by the IEA in Fig. 3.2.1 are also equivalent to the per capita use of 0.17 $toe_{EP}$·(cap·y)$^{-1}$. The agency statisticians probably used above model to include basic and unreported requirements. The production of biofuels was negligible in 1973 and amounted to 52 $Mtoe_{EP}$ in 2009 [20]. On the other hand peat is aggregated with coal in the IEA convention, but its contribution is small.

For the basic tasks of agriculture, before the industrial revolution and continuing a long time afterwards, humans had to rely on muscular work supplied by themselves and/or animals. For some processes such as grain grinding ancient societies also used water power or wind mills. But their works did not represent a large amount of energy and could be done by animals. Muscular work requires just more food. This form of energy is not accounted as such by the reporting agencies. Assuming that the high heat value of digestible part of the food per capita and per day is 3 000 kcal·(cap·day)$^{-1}$, which provides for the human metabolism and some physical activities, it amounts to 0.11 $toe_{cal}$·(cap·y)$^{-1}$. Under stresses - overpopulation, adverse weather...- this amount can be decreased, but not that much. Food for draft and working animals should also be included. As the cultivated area dedicated to their food can be larger than that for humans, their contribution is at least equivalent to the requirements for humans. These overall food needs are of the same order of the basic energy requirements provided by wood based fuels and wastes. Both are governed by the same limiting factors, agriculture or forestry yields and available surfaces. Human density and consequently population are thus limited by these factors in absence of other energy.

According to rough estimations of work output [31], muscular energy per human varies from 0.009 to 0.03 $toe_{mus}\cdot(cap\cdot y)^{-1}$. It gives a rough base to compare with the consumption of final energy like electricity.

## Analysis of the Correlations - Mechanisms of the Two Areas of Growth

The correlations shown in Table 3.2.3 do not give a clear indication of the mechanisms underlying them. Which growth is responsible for the other?

A growth in energy consumption corresponds to a growth of production, obviously. How does this latter growth occur? By discovering natural resources rich in energy? The existence and use of bitumen were known since antiquity. As early as the thirteenth century, due to woods disappearance through overexploitation in some developed regions (Holland, Wallonia, Great Britain [2], the Loire basin in France [32]...) peat and coal at the surface or at shallow depths were substituted for wood as a fuel.

But their use was very limited. At best these early exploitations resulted in localized prosperity. Their increase usage depended on better or entirely new processes to extract a large amount of resources, to transport it and thus to deliver their heat value to the end users.

The steam engine responded to all these needs. It was first developed successfully in Great Britain in 1712 by Thomas Newcomen to make water filled coal mines accessible by pumping excess water and thus increase the coal production [1, 2]. The possibility to use steam for work was already envisioned much earlier in the antiquity. It did require an accumulation of knowledges, first technical then fundamental to attain practical and large scale application. Past misconceptions had to be corrected. T. Newcomen was probably aware of the previous experiments like that of D. Papin who used cylinder and piston. Newcomen's machine required some metallurgic processing, even rough, of iron and other metals to manufacture components like the boilers, cylinders, piston, rods, pipes and chains, only possible at his life time. Nevertheless, he had to work out a lot of crucial details, like the seal between the cylinder and the piston, with his limited means. His engine converted only a mere 0.3% of the coal heat to work (see the section 5.1.1 of the chapter 5).

It took at least 60 years to use a separate condenser to improve the engines efficiency. First elements of thermodynamic theory were established only in 1850, based on the works of S. Carnot and experiments of J. Joule. They provided engineers with powerful quantitative tools to improve the steam engine, conceive new engines and extrapolate output at larger scales. All of these machines lead to an increase of exploitation of coal in mines, and the outputs of more modern industries with higher productivity.

This development was also a response to demand pressure arising from a lack of wood fuel and the population increase. The first steam engine was purely intended to lift water from deep mines and extract more coal. But its (relative) success did not materialize before the 18th century and before sufficient applied knowledge had accumulated in spite of earlier use of coal.

Similarly, the industrial exploitation of crude oil was conditional of the refining technology. These depended on progress in both theories and industrial application of chemical processes.

Would it otherwise be possible to have population growth without a prior increase of energy production?

Access to food, and so higher agriculture yields in t per ha.year and/or larger cultivated area, are the first drivers to a growth of population. Food also provides basic energy for muscular works. In the early Middle Age (roughly from the year 1000 to the beginning of the 14th century) the population in the area corresponding to present day France grew rapidly to reach about 20 millions capita [33]. But this rise overcame the agriculture output limits fixed by the techniques providing sufficient nutrients to the soils [6]. Agriculture, and then demography, became very weather dependent. With a cooler climate after 1300 and lower yields, the population fell to 10 millions of people around 1430 and did not surpass the 20 million mark again until the 18th century (population also suffered from diseases, like the Black Death around 1350, and extended wars). Improved agriculture practices after 1700 and also better knowledge of nutrient role allowed the slow increase of agricultural yields.

We can observe a common pattern between growth of food production and energy production, both occurring through improved or new processes due to better knowledge. Moreover, with the advent of synthetic fertilizers to increase yields, a higher production of food required larger inputs of energy. Furthermore, as seen above, humans additionally require some basic forms of energy for their heating and cooking, as well as to manufacture materials such as glass, bricks, quicklime...for their buildings. Other factors like transportation, access to clean water and better sanitation played an important role in population growth. However, their achievement took place largely after the 19th century and all required abundant and cheap energy.

The overall conclusion seems that population growth is driven by the progress in technologies of large production of useful and cheap energy, which is not dependent on the population size but on the sufficient accumulation of knowledges. This emphasizes the importance of the energy efficiency resulting from the works of J. Watt and S. Carnot on the steam engine.

Nevertheless, there is a positive feedback between population and energy production. Table 3.2.3 shows that without constraints on the production, which translates to low prices of energy, the human tendency is to increase its comfort and the energy consumption associated with it.

It should however be noted that with the comfort and the security provided by abundant energy consumption, the population tends to stabilize, reducing the number of births after a demographic transition.

### 3.2.2. Energy Efficiency and Energy Conservation

At the and user level, or close to it, energy can be saved to partly offset the overall increase of consumption due to the population increase and its overall wellbeing. We must be careful to distinguish between the two contributing elements of these savings, the energy efficiency of industrial processes and energy conservation. The latter depends on the attitude of the consumer especially at the household and commercial levels.

If we look at one of the most basic needs, food, and at the energy to produce it, both contributions to savings are present. Gas requirements to produce proteins in plants *via* the synthesis of N fertilizer can be minimized at the factory and at the farm (section 3.1.1). However, increasing consumption of animal proteins per person, even in excess of the physiologic need, will void all previous efforts of savings.

In the case of another basic need - that of space heating -, efficiency corresponds to improvements in converting a fuel into useful heat, and to reduce the heat need per unit of floor area and per unit of indoor/outdoor temperature difference. It implies lower conversion losses in the heater and the use of better thermal insulation for the building. But the residents can erase these efficiencies by increasing their living areas or dialing up the inside temperature. The consumption rate in $J_{th}$ per unit of house area to heat and per year, that is readily available, does not distinguish between the two contributing factors. The local value of heating degree-day - the difference between inside and outside temperatures when positive integrated over a year - must be known and its influence removed.

The same observation can be made about TV panels and the massive introduction of flat liquid crystal displays in recent years (see the section 4.2 in chapter 4). Their consumption in use per $cm^2$ has decreased, but the surface per panel has increased and they remain switched on for longer times.

The distinction between efficiency and conservation can be illustrated in the times before and during the first period of the industrial revolution. Industrial processes were then inefficient (low yield of Newcomen steam engines and early production of ammonia - see Fig. 3.1.1 -). But low quantities of the output were consumed - in total and per capita. That situation resulted first from forced conservation through poverty of the large majority of the population.

The distinction implies the existence of an opposite trend to the one as currently observed. We can hope to have both positive effects resulting from energy efficiency of the industrial processes as well as energy conservation by the consumers (with the help of taxes).

**Developments of Gasoline Consumption and Travel in the USA from 1980 to 2011**

The end user behavior depends much on the price he/she has to pay for his/her energy purchases, as well as on the long term investment required to reduce this consumption, both relative to his/her income. Consumption trends or fashions have also an influence.

These consumer choices, like the car size and its resultant mass, are more relevant to the human behavior than to the industrial process efficiency. Efficiency deals with the yield of the internal combustion engines and with the reduction of air and wheel drags. Improving the average yield of the vehicle engines under very different conditions, from the highway drive to everyday travels over short distances in congested cities, is getting ever more difficult after one hundred years of research (cylinder reduction with higher injection pressures, lowering frictions...). Moreover the means to limit their contamination reverse the gains (back pressure due to filter for finer particulates, consumptions to regenerate the filter and catalytic converter, processes to reduce NOx...). Smoother car design and better tires result in too modest gains of efficiency.

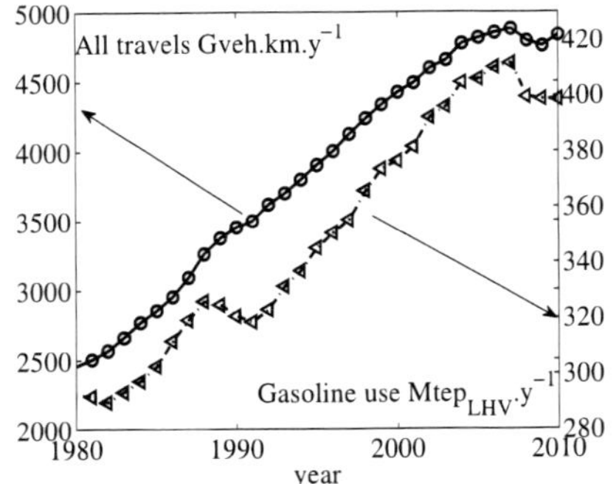


(a) Annual distance traveled by all vehicles (o continuous line, left axis) and gasoline consumption - including ethanol - (◁ dashed line, right axis) in the USA from 1980 to 2010.

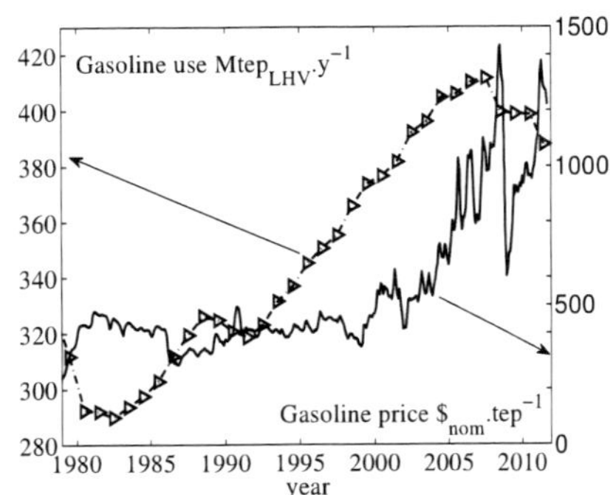


(b) Annual consumption (▷ dashed line, left axis) and monthly price in nominal $\$.toe_{LHV}^{-1}$ (continuous line, right axis) of gasoline from 1978 to 2011.

Figure 3.2.3. Influence of the gasoline price on consumption and distance traveled. Sources: Federal Highway Administration 2011 [34], US EIA [19].

Figs. 3.2.3 and 3.2.4 show how these factors are in play in the case of the residential transport in the USA between 1980 and 2011. The overall distance traveled by all road vehicles in the USA increased continuously from 1980 to 2007 (Fig. 3.2.3(a); the indicator includes also the diesel fueled vehicles assumed to follow the same trend). The number of all vehicles has risen even more rapidly than the population (Fig. 3.2.4(a)). Average distance per vehicle increased to 20 000 km until the end of 90's.

The motor gasoline consumption has risen since 1982, except around 1990. The shock of oil prices at the end of 1979 had had two effects (Fig. 3.2.3(b)): the gasoline consumption was reduced by more than 10% over a year, and the sales of low fuel consuming cars were boosted (table 4.7 in [34]). The latter effect was felt later at the end of 80's in spite of increasing distances (Fig. 3.2.3(a)). But after 1985 and the fall of the crude oil price, lower fuel consumption of a car was not anymore a sale argument. Sales of large and heavy cars with a higher consumption per km increased (Tables 4.7 and 4.9 in [34]).

The fashion for big cars was such that the progressive increase by twofold and half of the gasoline prices from 2000 to 2006 - with some fluctuations - was not effective in reversing this trend (Fig. 3.2.3(b)). The rapid rise of prices in 2007/2008 provided the triggering shock that did influence the motorist be-

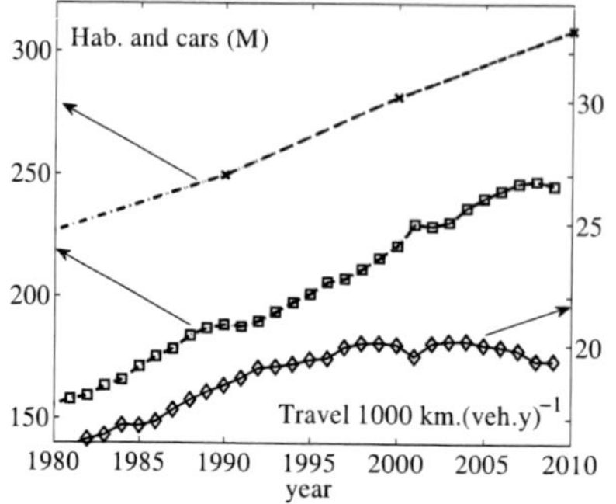


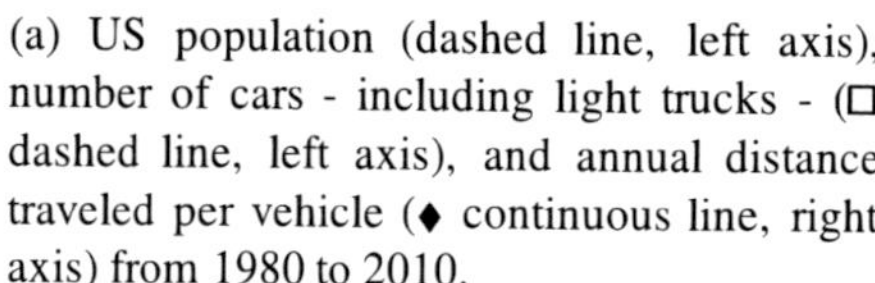
(a) US population (dashed line, left axis), number of cars - including light trucks - (□ dashed line, left axis), and annual distance traveled per vehicle (♦ continuous line, right axis) from 1980 to 2010.

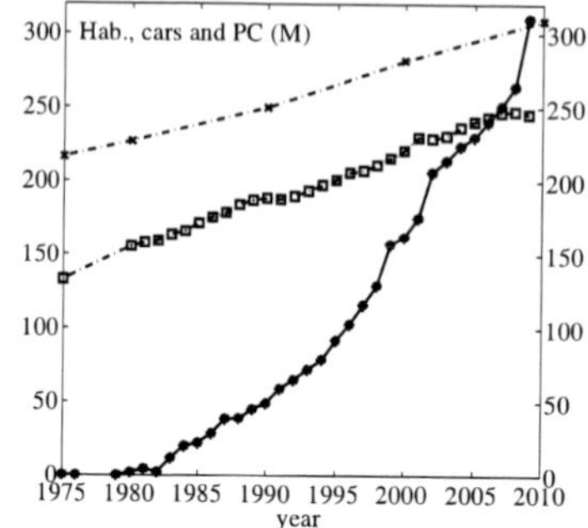


(b) US population (dashed line, left axis), number of cars (□ dashed line, left axis), and number of personal computers in millions (• with continuous line, left axis) from 1980 to 2010.

Figure 3.2.4. Evolution of population, number of vehicles, distance traveled per vehicle and number of personal computers in the USA from 1980 to 2010. Sources: US census bureau 2011, Polk company, Federal Highway Administration 2011 [34] and Computer Industry Almanach Inc. press news.

havior. Consumption decreased with the reduction of travel distances and the number of vehicles in circulation. However, the decrease reached a modest 3% over the year. Relative to the gasoline price at the dispenser, the American driver had comparatively higher purchasing power than in 1980.

The same situation arose and reactions occurred in 2010/2011. The low sensitivity of consumers to gasoline price in the short-term can be explained by the significance of the fixed costs (car maintenance, insurance...), as well as that of the investment to buy a new and more fuel efficient car. Similar to what occurred between 1980 and 1990, the shock effect will probably be felt in the long-term.

The last figure shows the rapid rise of the number of personal computers in the USA (Fig. 3.2.4(b)). So far they have had a weak influence on the number of cars and the travel distance, although teleworking has the potential to reduce them. They have added new applications and a new form of consumption. Computers contributed to the increase of electricity demand per capita in the USA observed between 2002 and 2007 (see Fig. 3.1.2). In the future, with behavior changes and thanks to the Internet access, this technology could play a major role in reducing the physical travel requirements.

### 3.2.3. Summary of the Role of Population

A human being requires at least about 0.15 to 0.20 $\text{toe}_{\text{EP}}\cdot(\text{cap}\cdot\text{y})^{-1}$ of energy for his heating, cooking, fabrication of basic shelters and tools usually in the form of biomass and waste fuels. If abundant resources and efficient processes to extract and make them useful exist, he/she will increase his/her requirements of useful energy. Population will also rise, especially from a base situation where its requirements were initially constrained.

Consequently the consumption at a country or at the world level increases with the demography and the consumer tendency to seek higher comfort. Thanks to these drivers and above all progresses on processes (one of which energy efficiency), the last century experienced the highest consumption and growth levels in all of the human history.

What would be the limits of these growths?

**Table 3.2.4. Three projections of population and consumptions of primary energy and final electricity until 2050**

| | 2009 | | 2020 | | 2030 | | 2050 | |
|---|---|---|---|---|---|---|---|---|
| Gcap | 6.8 | | 7.7 | | 8.3 | | 9.3 | |
| Gtep | prim. | elc | prim. | elc | prim. | elc | prim. | elc |
| 1st proj. | 12.2 | 1.44 | 15 | 2.1 | 18 | 2.9 | 27 | 5.6 |
| 2nd proj. | 12.2 | 1.44 | 15 | 2.0 | 17 | 2.6 | 22 | 4.3 |
| 3rd proj. | 12.2 | 1.44 | 19 | 2.7 | 25 | 3.8 | 37 | 6.0 |

Table 3.2.4 provides potential developments of both population and energy consumption towards 2050.

As far as the population is concerned, projection was issued by the organization of the United Nations in its median scenario based on current fertility rates (at the beginning of 2013). Population tends to stabilize. The trend seems not directly related to energy abundance but as a consequence of a better standard of life, especially for women.

Three methods are used to forecast future energy consumptions, giving three different results. The first one assumes the same growth rates of consumptions as those for the last period until 2009 according to data in Table 3.2.2. The second method is similar but working with per capita consumptions of Table 3.2.3. The US per capita consumptions are, however, assumed to remain constant after 2009 because of its already high level.

The last and third scenario postulates that all the humans would reach in 2050 the level of life standard of a French in 2009 - and subsequently per capita consumptions (Table 3.2.3) -, resulting in a overall consumption of primary energy close to 40 $\mathrm{Gtoe_{EP}{\cdot}y^{-1}}$. From an amount of 12.2 $\mathrm{Gtoe_{EP}{\cdot}y^{-1}}$ in 2009, it would require a constant growth of 3% per year. This rate is higher than the rate since 1973 - 2% -, but lower than the one during the prosperous period of 60s and 70s - 6% -.

All scenarios assume that primary energy would be still flowing abundantly at reasonable cost, and thus supply would not be a limiting factor.

The extrapolation of the growth of per capita consumption over the period between 1973 and 2009 gives the lowest requirement in 2050, 22 $\mathrm{Gtoe_{EP}{\cdot}y^{-1}}$. Because of the importance of population in the consumption the scenario is probably, and hopefully, the most plausible one.

What can be said about the resource availability to achieve this growth?

## 3.3. Abundance of Energy Resources and Efficiency of Processes

What about the resources to sustain the energy requirements of the economy and the human population for basic life necessities and discretionary use? What types of resource are available to us and which of these are actually accessible? What are the limitations to their exploitation? And how does the efficiency of extraction and conversion processes play a role in their use?

The efficiency of processes to extract a resource and transform it into a useful energy or material provides no information on the abundance of the resource. The processes can be efficient but the reserve poor. However, the amount of useful energy which can be produced from the resource will depend on this efficiency.

We distinguish between the energy resources and the material resources such as building materials, metal ores... [35]. The latter ones require the former and can be recycled if there is enough energy (for instance the treatment of salt or contaminated water to obtain fresh water). The energy resource is thus

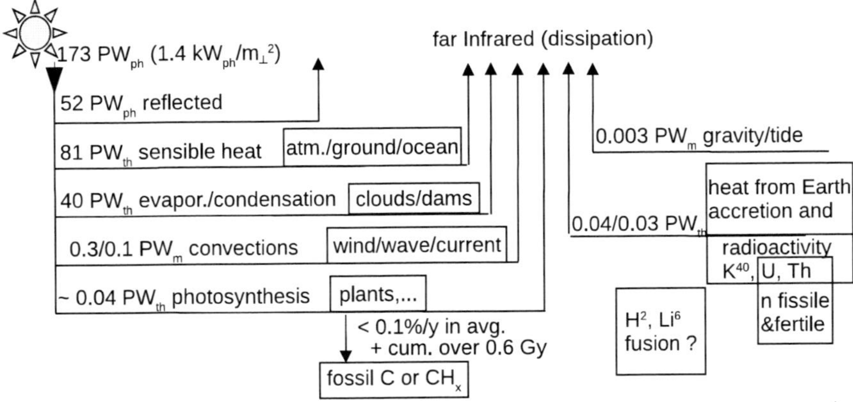


Figure 3.3.1. Flows in PW ($10^{15}$ W) of natural resources with energy content at Earth surface [36, 37]. Subscripts of units are **m** for mechanical, **ph** for photon and **th** for thermal. Boxes represent natural or human made storages where useful energy can be drawn from. Isotopes of U and Th inside the Earth can provide energy by fission or otherwise decay naturally with the release of heat. $Li^6$ and $H^2$ might be used in fusion reactors. By comparison primary energy consumed by humans amounts to 12.2 $Gtoe_{EP}$/y or 0.016 $PW_{EP}$.

considered the limiting factor for our industrial activities.

## 3.3.1. Energy Resources Close to Earth Surface

### Energies of Flow

Fig. 3.3.1 shows the flows of various natural energy at the Earth surface. They are huge, especially the solar one, in comparison with the flow of primary energies consumed by humanity. The solar flow reaching the surface represents around 90 $PW_{ph}$ - the rest, 85 $PW_{ph}$, being reflected or absorbed in the atmosphere - against 12.2 $Gtoe_{EP} \cdot y^{-1}$ or 0.016 $PW_{EP}$ (see Fig. 3.2.1). One way or another this flow is dissipated in the form of heat at temperatures of Earth surface or its atmosphere, and radiated continuously back into space. Before that final transformation the energy from the sun has been transiently stored on Earth for between few hours and more than a billion of years.

Contributions to heat flows other than solar radiations are very tiny by comparison. One is the dissipation of the mechanical energy resulting from the attraction of Moon and Sun on Earth, *i.e.* the gravity tide. Another results from the heat accumulated in the Earth depths, itself due to the decays of radio-nuclides and to the remainder of the gravitational energy released during Earth formation.

**Energies of Reservoir**

The existence of natural energy storages is also indicated in Fig. 3.3.1. These storages permit to regulate the flows of natural energy and constitute the origin of nearly all primary energies (solar radiation and tide are the only energies of flow used industrially, and still remain marginal). A large part are supplied by the continuous solar radiation after its transformation into other energy forms.

A fraction of radiation is thus absorbed in atmosphere, soil or ocean surfaces to form stored sensible heat (heat with temperature higher than the ambient one). Some of the part absorbed at the earth surface - and replenished by a solar flow close to 50 $PW_{ph}$ - can be used by humans. Stored heat presents residence times from one hour to a year.

Solar radiation is also converted into latent heat owing to water evaporation and transpiration (the global mass of water flow is thus about 16 $Tg \cdot s^{-1}$, using the water latent heat, 2.45 $MJ_{th} \cdot kg^{-1}$ of water). Heat is released during the condensation of the vapor into drops or crystals in atmosphere. The mechanism also provides the energy for lifting water in altitude. The water gravitational energy can be stored in human made dams before providing work to a turbine (according to the 2010 survey of energy resources by the world energy council, the global gravitational potential is estimated at 0.005 $PW_m$ from precipitations and water run-off). Average residence time of water in atmosphere is about ten days. Water amounts in dams are regulated for a cycle of one year at most.

A tiny fraction of the stored heat creates thermal convection at different scales, from Earth size (ocean currents) to few km (sea or ground air breezes). The wind energy is stored in atmosphere for few to several days. Wind resource estimated from air velocity measurements corresponds to the stored energy not the fraction of solar power converted into air convection shown in Fig. 3.3.1.

An even smaller fraction of solar radiation is converted through photosynthesis into biomass for few months to several years.

The most important forms of stored energy for humans are the organic residues from decayed plants accumulated slowly into the Earth crust over at least the last 600 millions of years. More than 80% of primary energy comes from this resource pool (Fig. 3.2.1).

Another potential form of stored energy in the Earth crust, which provides the humanity with close to 6% of its primary energy, resides in the nucleus of heavy atoms amendable to neutronic fission, isotopes of uranium (U235, U238) and thorium Th. Their life time is about few billions of years, equivalent to Earth age.

Their natural radioactive decay, as well as that of their daughters and of potassium 40, also supplies part of the geothermal heat stored in the Earth depth. The heat temperature, and consequently its quality, increase with the depth.

The disadvantage of the energies of reservoir is the risk of their depletion under a too large rate of draw down. By comparison with the human production - 16 $TW_{EP}$ -, we can distinguish among these energies (from flow values in Fig. 3.3.1):

- those with a natural rate of replenishment higher than the human production, such as sensible heat and wind from the sun,
- those with a close rate on average, such as water run-off, biomass and geothermal heat,
- and those with a lower rate, such as fossil organic material and heavy atoms (U, Th).

The flow of organic matter preserved in Earth crust represents on average less than 0.1 % of the yearly biomass production [38, 39], or lower than 40 $GW_{th}$ in terms of energy flow. In terms of mass of organic C the flow represents roughly 30 $Mt_C{\cdot}y^{-1}$ (the dried and ash free organic matter has 50% in mass of C and its heat value is close to 20 $GJ_{HHV}$ per $t_{DM}^{-1}$ of dry matter). This flow is not geographically and through geological times evenly distributed.

Most of the atoms in Earth were formed before the Earth birth - more than 4.5 Gyears - and no atom has been created since, except from natural radioactive decays of radionuclides such as U and Th atoms. Furthermore, due to the radioactive decay itself, the number of useful nuclei for the nuclear industry has been decreasing, especially U235.

Compared with these low rates of replenishment we can counter that the resources in place are extremely large. According to J. Hunt [39,40], the amount of organic C from decayed plants stored in the sedimentary rocks is of the order of 12 500 $Tt_C$ (to which may be added some elemental C - such as graphite - in metamorphic rocks, close to 3 500 $Tt_C$). This mass is equivalent to an energy of about 10 000 $Ttoe_{EP}$, when considering the heat value of C (organic matter also contains H in reduced form and consequently the amount is higher). The inventory is deduced from the analysis of the organic matter in tens of thousands of rock samples. For each type of sedimentary rocks, shale, claystone, carbonates and sandstone, its average content of organic C and its total mass in the crust are determined. The total mass of sedimentary rocks represents 2 250 000 Tt, or 8% of the total mass of the crust (28 500 000 Tt).

Another example is the U resource in the Earth crust. From the compositional analysis of rock samplings, the average U grade $C_U$ is 2.7 $g_U \cdot t_{crust}^{-1}$ corresponding to an amount of 75 $Tt_U$ in the entire crust [41]. This quantity can produce around 800 000 $Ttoe_{EP}$ of heat in a thermal neutron reactor, or 140 000 000 $Ttoe_{EP}$ in a fast neutron reactor. This is equivalent to a reserve of more than sixty millions of years or more than ten billions of years respectively, at the current consumption of primary energy.

Therefore both flow and reservoir energies could provide for the human requirements - even at 40 $Gtoe_{EP} \cdot y^{-1}$ according to the most optimistic of the scenarios proposed in section 3.2.3 - for much longer than historic times. However, a large resource in place is not a sufficient pre-condition to produce a useful energy.

### 3.3.2. Limiting Role of Industrial Processes and Their Efficiency

#### Fundamental Limitations

**Earth Surface Limitation for Fossil and Nuclear Fuels** A first theoretical limitation arises from the elevation of surface temperature $T_{surf}$ to radiate the heat dissipated by humans. Indeed, if we assume a constant growth in consumption of primary energy and that nearly all of it is supplied by the fossil fuels from the crust, their dissipation increases the heat to be radiated outwards. According to the Stephan-Boltzmann law about the total radiation of a black body with the temperature $T_{surf}$ in K [41], the heat flow increases with the fourth power of $T_{surf}$. Presently, to radiate the flow of dissipated energy from the sun - 120 $PW_{ph}$ -, $T_{surf}$ is on average 15°C or 288 K (also including the effect of the greenhouse effect of the Earth atmosphere). A yearly growth of 2%/y in energy consumption, extrapolating the trend observed since 1973 (section 3.2.1), would produce in 450 years an amount equivalent to 120 $PW_{ph}$, or 91 $Ttoe_{EP} \cdot y^{-1}$, which would double the heat to be radiated from the Earth. The temperature will have to increase to 345 K or 72°C. Life on Earth would be greatly changed.

But even economically such growth does not seem plausible. We can take a more reasonable value for the future consumption, stabilized at 40 $Gtoe_{EP} \cdot y^{-1}$ or 0.053 $PW_{ph}$. The temperature $T_{surf}$ will increase by a mere 0.5°C, which is not something to worry about too much.

**Air Oxygen Limitation for Fossil Fuels** Another fundamental consideration when tapping fossilized organic residues for our need, is the consumption of $O_2$ from the atmosphere. Indeed, the burning of the fossil fuels reverse the overall reaction of photosynthesis where $CO_2$ provides organic C + $O_2$. To synthesize 12 500 $Tt_C$ of organic C, an amount of 33 500 $Tt_O$ of $O_2$ was released, according to the stoechiometry of the reaction. A tiny part - 1 500 $Tt_O$ - remained in the atmosphere while the majority was stored in the crust in forms of iron oxides and sulfates [36]. A complete use of $O_2$ in the atmosphere would allow the combustion of 560 $Tt_C$ of organic C. An even bigger change for life on Earth than the rise of temperature $T_{surf}$.

If we restrict to a more reasonable use of 1% of $O_2$ in the atmosphere, only 5.6 $Tt_C$ can be used, which amounts to about 5.5 Ttoe of fossil fuels (assuming that the elemental composition of the organic matter in rocks is CH). It will offer less than 500 years of consumption at the current rate of 12 $Gtoe_{EP} \cdot y^{-1}$, or less than 150 years at a rate of 40 $Gtoe_{EP} \cdot y^{-1}$ for the high scenario of section 3.2.3.

On the other hand the exploitation of U or Th from the crust does not require oxygen from atmosphere (the nuclear energy is assumed to provide for all its requirements).

Both these fundamental limitations are avoided when using energies derived from the sun or geothermal heat.

However, more stringent limitations arise with the choice of available processes to extract the resources, taking into account their efficiency and potentially their economic constraints as seen in the first section of this chapter.

### Limitations of Processes

**Limitation to Extract the Crust Resources** How the huge resource of U or fossil organic matter in the crust could be processed?

Let's consider the open pit mine system. In the Rössing mine in Namibia the company Rio Tinto extracts U from a 300 m deep pit [42]. The main energy consuming operations are the excavation and the transportation of the material removed, and the grinding of ore in fine particles of few tenth of mm in order to extract from it the uranium. The energy consumptions are thus roughly proportional to the mass of processed ore, assuming a fixed strip ratio (the ratio of waste rock in overburden to ore). Hence the appropriate rate $r_{mine\_op}$ of consumption of final energies - electricity, fuels - for the open pit mining is expressed in $\mathrm{GJ_{EF}}$ per tonne of ore. Its value for the Rössing mines is about 0.1 $\mathrm{GJ_{EF} \cdot t_{ore}^{-1}}$ [42]. This order of magnitude is confirmed for gold and iron mines in Canada [43].

However, to study the impact of the mine operations in the nuclear energy chain and to compare it to the electricity production, we must convert the rate $r_{mine\_op}$ into another one, $R^{nuc}_{mine}$, representing the consumption of energy per tonne of U ($\mathrm{GJ_{EF} \cdot t_U^{-1}}$), or more interestingly per $\mathrm{J_e}$ of electricity produced at the nuclear power plant ($\mathrm{J_{EF} \cdot J_e^{-1}}$). The rate $R^{nuc}_{mine}$ represents the contribution of the mine to the overall rate $R^{nuc}$ defined to measure the efficiency of the entire nuclear industry. The conversion from $r_{mine\_op}$ into $R^{nuc}_{mine}$ requires the grade of U in the ore $C_U$ expressed in $\mathrm{t_U \cdot t_{ore}^{-1}}$ according to:

$$R^{nuc}_{mine} = \frac{r_{mine\_op}}{C_U}. \tag{3.3.1}$$

The conversion of U into electricity also requires knowing the yield of a thermal neutron reactor, which represents the main type of reactors operating in the world. Its yield of conversion is around 150000 $\mathrm{GJ_e \cdot t_U^{-1}}$.

In the case of the Rössing uranium mine, the grade $C_U$ is around 300 $\mathrm{g_U \cdot t_{ore}^{-1}}$. The higher U grade in the Rössing mine than the crust average was achieved thanks to a geological mechanism that concentrated U. The rate $R^{nuc}_{mine}$ in Rössing amounts thus to about 300 $\mathrm{GJ_{EF} \cdot t_U^{-1}}$, or 0.2% of the electricity at the outlet of the power plant.

What would be the energy consumptions to extract all the uranium of the crust? Assuming that the rate $r_{mine}$ for any mine remains identical to the rate $r_{mine\_op}$ for open pit mines, the mining contribution to the nuclear industry according to eq. 3.3.1 would amount to $R^{nuc}_{mine} \backsim 30000$ $\mathrm{GJ_{EF} \cdot t_U^{-1}}$, or 20% of the produced electricity.

The contribution to the overall rate is not anymore negligible. Moreover, because we can not limit mining to U extraction at the Earth surface, the rate $r_{mine}$ must at least take into account the energy to lift the removed rocks from depth and potentially to pump the underground water to access the rock. According to the data from the coal exploitation in the basin de la Loire (section 5.1.2 of chapter 5), the electric consumption to lift the ore and other rocks is close to the gravity energy taking into account a mechanical yield of about 80%. Hence for a depth $h$ in km, a total mass $M$ of rock and water to lift in t per tonne of ore, and by including the rate $r_{mine\_op}$ for the surface work, the total rate $r_{mine}$ expressed in $\mathrm{MJ_{EF}\cdot t_{ore}^{-1}}$ is:

$$r_{mine} = 100 + 12 \cdot M(\mathrm{t}) \cdot h(\mathrm{km}). \tag{3.3.2}$$

In the case of a mining yield $M = 2$ t of uplifted rock per ton of ore and a depth $h = 4$ km, the rate $r_{mine}$ would amount to about 200 $\mathrm{MJ_{EF}\cdot t_{ore}^{-1}}$, which is double the initial cost.

Moreover because of the regular rise of temperature and pressure with increasing depth (1 to 2 bar per 10 m), boring underground mines is getting more difficult, requiring more resistant materials. In fact no gallery exists below few kilometer depth.

As far as fossil organic matters are concerned, like U the vast majority is dispersed in the sedimentary rocks (average organic C content of about 0.55% as seen above) [39]. Its exploitation would require a rate close to the one in eq. 3.3.2. Even for shallower depth, and so without the gravity contribution on consumption, the rate $R^C_{mine}$ for a C content of 5 $\mathrm{kg_C\cdot t_{ore}^{-1}}$ amounts to 20 $\mathrm{GJ_{EF}\cdot t_C^{-1}}$, or about 0.5 $\mathrm{tep_{EF}\cdot t_C^{-1}}$. Considering that 1 $\mathrm{t_C}$ of C yields close to 1 $\mathrm{tep_{EP}}$ of primary energy of organic residues, or 0.5 $\mathrm{tep_{EF}}$ of final energies for the mine (a large part electricity), the whole system would be barely self-reliant.

Moreover, characteristics other than concentration and depth in the crust become critical. Processes to separate U from other elements after grinding, and their requirements, also depend on the type of minerals. Uraninite, the mineral most looked for, permits the use of relatively easy processes of separation with diluted sulfur acid, whereas monazite requires to heat it for several hours within 98% sulfuric acid at temperatures between 120 and 150°C, or hot and concentrated hydrofluoric acid). Fossil organic matters are overwhelmingly in an insoluble form, kerogen, which requires pyrolysis in a retort to obtain useful fuels.

**Recoverable Resources from the Earth Crust** Due to these limitations only a fraction of the resource, or reserve, can be extracted. The vast majority is too deep, too dispersed, and/or in unattractive chemical or physical states to be exploited efficiently with known processes. Limiting economic factors such as the labor and investment productivity reduce the recoverable fractions further. Hence for particular process and its resource, e.g. open pit mine for U extraction, there are only limited associated reserves suitable for extraction out of the total resource pool.

In the case of U open pit mine, the Rossing mine with its grade $C_U \simeq 300\ \mathrm{g_U \cdot t_{ore}^{-1}}$ represents the lowest limit of U extraction. The mine operator benefits of a favorable labor and fiscal environments, and part of the infrastructure investment has been amortized.

When the physical and technical parameters of the deposit degrade rapidly outside the reserve volume, and in absence of a new process or a spike of the market price of the final product (U concentrate in this case), once the reserve is exhausted the extraction stops. For smoother transition out of the reserve the exploitation may proceed depending on market conditions. The global rates $R_{mine}^{nuc}$ of energy, material and labor requirements will rise more or less according to eq. 3.3.1, assuming U grade the main factor. These effects have to be offset by better sale price. Through mechanisms similar to the ones described for the photovoltaic sector in the part 3.1.3 of section 3.1, the whole economy will have to bear the price rise when purchasing nuclear based electricity.

Processes of U extraction other than mining exist such as the efficient *in situ* leaching. U is recovered by injecting in the rock alkaline or acidic solutions to dissolve the U mineral (cutting down, transport and grinding operations are absent). It allows to work deposits with a grade as low as $C_U \simeq 100\ \mathrm{g_U \cdot t_{ore}^{-1}}$. However, the process requires very permeable rocks, usually sedimentary rocks where U has accumulated by reduction. Hence globally a limited reserve base is associated with this process, lower even than the reserve of the open pit process. Moreover, a new process requires a long time of trials before its development at industrial scale.

In all cases the thermodynamic laws would limit the reserve of an energy resource.

In the case of fossil fuels (coal, gas and oil), their recoverable quantity can be tentatively estimated in the order of 10 Tt (of which 0.6 Tt has been already extracted), mostly coal [37]. Incidentally the quantity is higher than the one fixed previously by limiting at 1% the air oxygen intake, close to 6 Tt of coal. It represents a tiny fraction of the huge resource 12500 $\mathrm{Tt_C}$ of organic matter. Coal was formed during geological times in subsidences where large quantity of terrestrial plants had accumulated (usually in humid tropical or equatorial climates). Hard coal with high energy density is obtained after a long natural pyrolysis above 150°C due to the progressive burial of fossils in the crust. Peat represents the first stage of this process at shallow depth. Coal or peat can form large and/or thick beds where minerals (ash) and residual water represent from less than 10% to 60% of the mass depending on the maturity. No thin grinding is performed at the mine processing plant (actually it is done at the power plants operating with pulverized coal).

Therefore coal can be extracted with open pit mining more efficiently than U concentrate when consumptions are expressed per unit of mass of the final product, either coal or U concentrate. On the other hand, coal energy density is much lower than the U one. Hence the rate $R_{mine}^{elc}$ in $J_{EF} \cdot J_e^{-1}$ is more pertinent to compare the contribution of the mine to the production of electricity between both primary energy.

Petroleum and asphalt are produced from a type of kerogen largely made of organisms encountered in lakes or seas. They require a long pyrolysis above 50°C obtained below a depth around 1 km in the crust. Temperatures above 160°C - or depth higher than 4 km - produce eventually carboneous residues and methane. The amount of petroleum and asphalt represents roughly 550 $Tt_C$ of organic C, mostly dispersed in fine grain sedimentary rocks [39, 40]. Where the kerogen was sufficiently concentrated, enough amounts of hydrocarbons can be generated to crack the rock. Recoverable hydrocarbons, crude oil and gases, are in form of a low viscous fluid trapped under pressure in a permeable and porous rock, after a migration from the tight source rock. Their amount in these reservoir rocks is about 1 $Tt_C$. A long chain of natural operations, each with a low yield, explains the overall very tiny yield from the dead organisms to the hydrocarbon in reservoir rocks. A relatively simple and cheap process to extract the hydrocarbon, by comparison with mining, consists of drilling a hole to the reservoir rock and of using the difference of pressure to move the oil from the rock to the hole and then to the surface. A roughly quantity of 0,5 $Tt_C$ could be thus extracted.

But when the oil is in the form of tar stuck in the rock, producers have to resort to mining technologies such as open pit mine, if at shallow depth, or to *in situ* heating to make the tar less viscous. In Athabasca region in Canada the asphalt is in unconsolidated sandstone at a content of more than 10%. Asphalt results from the partial degradation of the crude oil after its migration in reservoir rocks too close to the surface. Its extraction requires more energy and investment for the same output than the well process. However, the chemical form, the concentration, the depth and the nature of the rock make them still physically and even economically profitable. The latter constraint depends of course on the price paid for the end product.

**Energy from Constant Flows** Useful parts drawn from energies of flow, or of reservoirs with a large replenishment rate, are also constrained by the existing processes and their efficiency. Both can be determined by fundamental limitations.

Presently wind is exploited by large three blade wind turbines, the development of which took nearly one century (it occurred mostly in Denmark). They can draw wind energy only from the first two hundred meters of the atmosphere. To be useful the turbines require strong enough winds with small turbulence rate and low variation over time, like the wind encountered in a flat coast or offshore. As a result the higher is the annual output for a fixed power capacity of a turbine, the lower are the number and surface of sites where this yield can be obtained.

Moreover wind turbines in a park must be sufficiently spaced to avoid disturbing the wind flows up and downstream, resulting in only a few % of the energy of the incoming wind over the turbine park to be converted into electricity.

Offshore windmills can utilize a higher quality wind resource (larger period of stronger and more regular wind). However, due to their extensive anchoring foundations they require substantially more materials in forms of concrete and/or steel per output power capacity than onshore turbines. If we except the energy consumed by some auxiliaries components in the turbine (like motors to adjust the direction of blades...), the main consumption results from the manufacture of the turbines (see the next paragraph). It will be amortized by the total net electricity produced during the turbine life time.

As far as solar energy is concerned locations where periods of direct sunshine are much longer allow a better amortization of the panel investment in terms of energy, as seen in section 3.1.3. Energy production using solar concentration thanks to optics can only operate with direct radiation and a tracking system. This limitation reduces the places in the world where to apply the system, and consequently the available resource.

### Concentrated and Diluted Energies

**Examples of Energy Density of Resources** When comparing the exploitation of U and kerogen with the same concentration, for instance 1% in rocks, U appears physically and economically recoverable, not kerogen. From an unit of volume or mass of rocks both exploitation will not produce the same amount of useful energy. Hence the heat value density of the raw material is an important variable of the efficiency of an extraction operation at the industry level. The rate $r_{extrac}$ of energy consumptions to extract and process the rocks or ore at the mine would be roughly constant if reported per unit of ore volume or mass. However, the rate for the extraction operation which matters at the level of the overall industry *indus*, is the global rate $R^{indus}_{extrac}$ in J consumed per J produced. Hence $R^{indus}_{extrac}$ depends on the quantity of energy finally produced from the unit of the raw material extracted, or its heat value.

The following list provides the heat content per $m^3$ of various natural resources. It is expressed per unit of volume at ambient conditions rather than per unit of mass as the volume is more relevant to fix the size of turbines and pipes, and to deduce the consumptions of transport and storage (especially for gases versus liquid or solid):

- 1 $m^3$ of ore with 0.1% U grade gives in a neutron thermal reactor more than 1 TJ of heat,
- 1 $m^3$ of crude oil will produce less than 40 GJ of heat,
- 1 $m^3$ of natural gas at ambient conditions will produce about 40 MJ of heat,
- 1 $m^3$ of water fall over 100 m produce about 1 MJ of work,
- 1 $m^3$ of air at 25 $km.h^{-1}$ contains about 30 J of kinetic energy.

Because of the origin of its energy from the atom nucleus, instead of energy from the external electrons in the case of the chemical energy, uranium benefits from a huge energy concentration. Wind suffers from being a very diluted resource.

**Other Factors at Play: Wind Versus Fossil Fuels** However to quantify this effect we must take into account that the extraction processes are different between these forms of energy, and subsequently their local rates $r_{extrac}$ per unit of volume.

Besides, we must consider the overall chain of operations of the industry. The consumptions of some of them are independent of the energy density of the raw fuel.

Let's consider few examples using the most diluted energy, wind. The material requirements for the production of electricity from wind turbine are first examined relative to the requirements for the production of gas at a North Sea field.

The company Vestas indicates that its onshore wind turbine V80 of 2 MW capacity requires 231 t of steel for the tower and nacelle, 27 t of iron and 800 t of concrete for the foundation, and 34 t of epoxide and fiberglass for the blades. The steel and iron use amounts to about 130 t per MW capacity. A large part can be recycled (except probably one in the concrete for the foundation). The wind turbine is assumed to operate during twenty years (limited by fatigues due to vibrations) at an average output representing 25% of the capacity (average in Denmark, lower in Germany and France). As a result one $GWh_e$ of output from the wind turbine requires about 3 tonnes of steel.

The Frigg complex (it includes various fields) is located 230 km northwest of the city Stavanger between the continental shelves of the United Kingdom and Norwegian borders in North Sea (see the site of offshore-technology.com/projects/frigg-field/). The field was discovered in 1971 at a water depth of 100 m. It was shut down in 2004 after 27 years of production. A total of six installations were removed from the field, except the gravity base structures of three of the platforms made of concrete. The installations contained the equipment to extract and process oil and gas. About 85 000 t of steel were recovered and recycled. A total of 48 wells were plugged and abandoned. Their pipes and tubes were not recovered. Hence we estimate the overall requirement of steel close to 100 000 t.

In absolute terms the mass for gas exploitation is much larger than that for the wind turbine. However, in terms of rate of steel consumption the situation is different. The cumulated gas output of the complex over its lifetime was 192 $G{\cdot}m(st)^3$ of gas[3]. The heat value of 1 $m(st)^3$ of North Sea gas is 39.5 $MJ_{HHV}{\cdot}m(st)^{-1}$, or 36 $MJ_{LHV}{\cdot}m(st)^{-1}$ *i.e.* 10 $kWh_{LHV}{\cdot}m(st)^{-1}$, according to the International Energy Agency [29,44]. For a rigorous comparison between wind turbine and gas field we convert the gas energy into electricity using the yield of a combined cycle power plant (about 60% net/LHV). The total electricity production from the gas would amount to 1 150 $TWh_e$. Therefore one $GWh_e$ of output from the gas production at Frigg required less than 0.1 tonne of steel at the field site, most of it was recycled.

[3]st: standard conditions where the gas is at a temperature 15°C and a pressure 1013 hPa.

Now let's compare the material requirements for the production of electricity from the wind turbine - 3 tonnes of steel for one $GWh_e$ -, with the one for the production from thermal power plants.

A coal fueled power plant using steam turbines necessitates for its construction around 50 t of steel per $MW_e$ of nominal output power [45]. Close values are also obtained for a gas or a nuclear power plant. Thus the pressurized water reactor Sizewall B of net capacity 1 190 $MW_e$ required 65 000 t of steel, or 55 t of steel per $MW_e$. The amount of steel depends thus on the electrical or thermal capacity of the plant rather than the fuel used. The quantity is more related to the size and mass of turbines, which themselves are fixed by the heat density of the steam or burnt gases (see the part 5.1.1 of chapter 5). Hence the interest to operate at high pressure. As a result the steel requirements per MW of capacity are rather comparable between the turbine wind - 130 t of steel and iron per MW - and the thermal plants, and do not reflect the energy density of the resource.

Now the real output of a thermal power plant can be adjusted close to its capacity in annual average - up to 90% -. Moreover, the building and equipment last longer than for windmill, at least 40 years in the case of a water pressurized nuclear reactor like the reactor Sizewall B. With a utilization factor of 80% and a lifetime of 40 years, one $GWh_e$ of output from a thermal plant requires thus less than 0.2 tonnes of steel and iron, a large part also recycled.

New thermal power plants have even a better rate. Westinghouse pressurized water reactor AP1000 of net capacity 1 117 $MW_e$ requires about 12 000 t of steel, or 11 t of steel per $MW_e$. Moreover it is safer than the old reactor and has a warranted lifetime of 60 years instead of 40 years. At a utilization factor of 90%, the rate of steel requirement falls to 0.025 t per $GWh_e$.

To be complete on the comparison between both situation, the production of the thermal power plants can be maintained constant with some charge following, which fits better the demand at the scale of a country, whereas the electricity output from wind turbines remains variable and random even at a country scale. A perfect comparison implies to include the necessary infrastructure such as electricity storage for a wind turbine to produce an output comparable with that of a thermal power plant.

### 3.3.3. Conclusions about Abundance and Efficiency

As a first conclusion for this section, the abundance of an energy rich natural resource does not mean the abundance of the useful energy produced from it.

The real useful part of the resource, or reserve $Q_{EP}$, corresponds to deposits workable with an efficient chain of processes of an energy system, such as crude oil in good reservoir rocks, or sites with regular, smooth and strong wind. These deposits present some characteristics like concentration, accessibility, state ..., which make them more likely to be recovered by comparison with the main part of the resource.

With the advents since mid nineteenth century of efficient processes to discover, extract, refine and use it, and because of its interest for the economy and humanity, crude oil from conventional reservoir rocks has become the dominant primary energy. Roughly, at the scale of a country like the USA, its production grew at an exponential rate [46]. But due to its limited quantity in reservoirs and the rapid degradation of content outside the reserve volume, the production must mathematically tend to zero after reaching a maximum, as described in 1956 by M. K. Hubbert. He predicted the amount $Q_{EP}$ of crude oil reserve in the USA outside Alaska - from 20 to 27 Gt -, and hence the year of the maximum of production, within an interval of error. The current value of $Q_{EP}$ for crude oil in the USA outside Alaska is estimated at about 30 Gt[4].

M. K. Hubbert was less lucky with the world production of oil and other energies of reservoir such as gas and uranium because his model relies on a good estimation of the total recoverable quantity $Q_{EP}$ of the resource, which is not always available at an early stage of the production. Nevertheless, the evolution over the time of the production of these limited deposits will present this peak or bell like form, albeit with a characteristic length of time somewhat larger than predicted.

The notion of reserve $Q_{EP}$ also applies for the energies of flow like the energies derived from the sun. But, in this case, the reserves are defined as a flow or power limit, and not as a store of energy. After a period of growth their production will reach a plateau fixed by the limiting flow $Q_{EP}$.

The reserve of energy $Q_{EP}$ and the efficiency of the energy system to process it are interdependent. Generally the better are the characteristics of the energy storage or flow, the smaller is the rate of consumption to extract and transform it into a useful form for human activities, but the lower is the associated reserve $Q_{EP}$.

If we define the rate of dissipation $R$ of the energy system according to eq.

[4]or 220 Gb, the main part already produced. The quantity is higher than Hubbert'forecast thanks to deep offshore deposits not anticipated at his time

2.2.1 in the section 2.2.2 of chapter 2 (the system is assumed self-reliant), the useful quantity $Q_D$ produced from the reserve $Q_{EP}$ is determined according to:

$$Q_D = Q_{EP} \cdot (1 - R) \tag{3.3.3}$$

Hence, when the rate $R$ reaches $R = 1$, the system is unable to produce an output, as shown for the dispersed kerogen in the earth crust. The equation presents thus the necessary condition for a working energy system. Value of rate $R$ is determined by the choices of processes and their variables. These choices depend on the reserve characteristics as seen above. However, they will be further limited by including economical, environmental and safety constraints (see section 2.3.1 of chapter 2). The respect of all of them can lead to a rate $R$ close to $R = 1$ or to a small reserve $Q_{EP}$. Humanity need would not be satisfied.

# Chapter 4

# Methodology to Determine the Efficiency of a System and Its Uncertainty

In the chapter 2 definitions of the efficiency of a system have been given in form of rates of energy consumption. Once the definition of a rate $R$ for the system is chosen we face the difficulty to estimate it due to the risks or error, the uncertainties of data and its sensitivity to the system parameters.

This chapter describes through examples a methodology and its mathematical tools, chiefly the global and local rates, to facilitate the study. Some applications of it have been given in the chapter 3 with the examples of the photovoltaic sector and the exploitation of resources in mines.

The necessity of a methodology arises from the complexity of any system, which means that the question of efficiency is at the system level but the answer is at the level of its processes. It implies some decompositions of the system towards simpler and independent ones.

The methodology cannot substitute for the raw data. But it points to the parameters and variables for which information is required and quantifies their importance. It permits to construct from the data the rate $R$ and its variables, within reasonable and controlled uncertainty and without too much time of analysis.

## 4.1. The Steam Engine. Risk of Large Errors

The methodology rests firstly on gathering enough data with sufficient details to avoid large errors. A sane start is to consider a unique value of the rate for a system unlikely.

An example of this caution is provided by S. Carnot himself in the last two pages of his book about the performances of the various steam engines known in 1824 [4]. He reported the very raw data in English units (bushel of coal, pounds of water and feet of height) before conversion into metric units. He indicated which machine was concerned, where and when. He mentioned an engine with two different cylinders of steam expansion of which the biggest had a 45 inch diameter and 7 feet stroke. The machine known as the Woolf engine operated with 3 bars steam, at the difference of the atmospheric Watt engines. It was working in a tin and copper mine called Wheal Abraham in Cornwall and was able to lift 195 $m^3$ of water over a 1 m height per kg of coal consumed (of which heat value was 7 $kcal \cdot kg^{-1}$, probably anthracite). The direct and practical rate of eq. 2.1.1 gives $R_D = 5.1$ $g \cdot (m^3 \cdot m_h)^{-1}$. But this result was obtained for a short period of one month. A more representative performance would be 104 $m^3$ of water over a 1 m per kg of consumed coal, or $R_D = 9.6$ $g \cdot (m^3 \cdot m_h)^{-1}$.

The best engines of Watt design could sometimes reach this yield. In contrast the old engine of Chaillot (the first steam engine in Paris erected in 1781 [3]) had a performance of 22 $m^3$ of water lifted over a 1 m per kg of coal, or $R_D = 45$ $g \cdot (m^3 \cdot m_h)^{-1}$. Therefore, if the various engines are not distinguished, a large error means here a factor five to ten. The engine of Chaillot represented the first type of Watt engines using a separate condenser but operating with a simple effect of steam injection [3]. However, it was able to save 75% of coal for the same work by comparison with the previous engines, the Newcomen ones, as claimed by J. Watt and M. Boulton at the start of their association. Consequently the yield of a Newcomen engine was 5.5 $m^3$ of water over a 1 m per kg of consumed coal, or $R_D = 180$ $g \cdot (m^3 \cdot m_h)^{-1}$.

An error of factor 35 is thus possible without details on the various early steam engines.

A steam engine in a present power plant, using turbines instead of cylinders for a better steam expansion, can lift 1 400 $m^3$ of water over 1 m from the heat value of 1 kg of coal, or $R_D = 0.70$ $g \cdot (m^3 \cdot m_h)^{-1}$. For a same generic system, the steam engine here, an extreme factor of error can reach 250. Obviously such confusion would not take into account all the progresses achieved over two hundred and fifty years, as well as the difference of thermodynamic conditions of the steam between the engines.

To understand these differences and to determine the variables which influence the rate, the steam engine must be considered as a complex system instead of a simple process. It was part of the Carnot's work to detail the successive operations of the thermodynamic cycle modeling the machine. A follow-up of the steam machine in chapter 5 gives some details on the theory and on the engineering, which permit to draw some predictions on future gains.

The following example of the phone manufacture presents the necessity of the decomposition of a complex system and its overall rate, as well as the method to carry it out.

## 4.2. How Much Energy to Manufacture a Phone? Global and Local Rates

The methodology exposed in this chapter shares many common features with the Life Cycle Analysis LCA. Thus, according to the ISO 14040 norm about LCA, the study requires the definitions of the system boundary and of its functional unit (equivalent to the global rate), it must rest on the raw data at the level of processes, use fundamental and technical information, carry out estimations of the accuracy... However, LCA is dedicated to identify the total environmental impact of a product through various indicators, such as the emission rate of various contaminants, that of fossil $CO_2$, the use of exhaustible resources... To examine how much a product impacts the environment, it is necessary to account for all the inputs and outputs throughout the life cycle of that product, from its birth, including design, raw material extraction, material production, part production, and assembly, through its use, and final disposal.

By comparison with the LCA objectives, the present study restricts mainly to the energy analysis. On the other hand, the rates and the methodology can be used to study other impacts than energy consumption as shown in the section 2.3.2 of chapter 2. Moreover the detailed study of an energy balance provides data to derive other indicators such as the different types and amounts of primary energy used, and hence $CO_2$ emission rate. The study has also to take into account the environmental constraints to the choice of the processes, as stressed in the chapter 2.

However, due to the complexity of each system it seems difficult to carry out all the environmental goals. Furthermore, because of their usually tiny outputs by comparison with the main output (otherwise it is a sign of mis-functions), estimates of emission rates of contaminants are even more uncertain and sensitive on process choice and their variables than the energy consumptions.

Besides, the LCA does not exploit sufficiently the possibility to adjust the depth of decomposition and to use local functional units, or local rates, as shown thereafter. hence it neglects established and specialized knowledges on processes, and hence incurs the risk of large errors.

### 4.2.1. Decomposition of a System and Its Rate. Local Rate of Its Parts

Figure 4.2.1. Photo of the Internet Protocol phone Cisco 7940G under study.

The case exposed here establishes and analyses the energy consumptions to make an Internet based phone, or IP phone (see photo in Fig. 4.2.1), using data from a commercial software based on LCA. The study was carried out during a student internship at the end of 2008 in the R&D laboratory of France Télécom (report available upon request). The French telecommunication operator bought the LCA software from a company dedicated to this type of analysis. As our purpose is not to denounce a person or a society but to show the difficulty of the work, its name is not disclosed.

**Decomposition of $R^{syst}$** The phone is the functional unit, or the product of the overall chain of fabrication. This chain is summed up by the diagram of Fig. 2.1.1 in chapter 2. The phone manufacture represents a complex system of which overall output is $L_{out}$, while its consumption is input $E$, either $E_D$ or

Figure 4.2.2. One of the components of the phone after its dismantling to achieve the Life Cycle Analysis.

$E_{EP}$, which is not relevant at this stage (however important when comparing data from various sources).

The LCA database does not contain directly the flows of the phone manufacture. However, it has information on the consumptions of primary energy to manufacture simpler industrial components, especially electronics, and materials, some of which are used in the phone. Hence a phone manufacture consists of the operations $j$ to make each of its components.

The consumptions $E_j$ of each operation $j$ are required to produce the final output $L_{out}$. Consequently the overall consumption $E$ is the sum of each $E_j$. Hence to deal with the difficulty of determining the rate $R^{syst}$ of a complex system, it is broken down into the rates $R_j^{syst}$ of each of its operations:

$$R^{syst} = \sum_j R_j^{syst}. \tag{4.2.1}$$

The choice of a yield indicator such as $Y = L_{out}/E$ instead of the rate $R$ would make difficult the decomposition in additive terms.

In the case of the phone, $R^{phon} = \sum_j R_j^{phon}$. The rate $R_j^{phon}$ is shorten in $R_j$ thereafter. The first part of the student internship consisted of dismantling the phone into each of its components, more than 80 (Fig. 4.2.2), and their identification to one of the items in the database, or to a similar one.

**Local Rates $r_j$ of the System/Phone Components** The operation $j$ itself is viewed as an independent system with a flow diagram similar to that of Fig. 2.1.1 in chapter 2. It requires the consumption $E_j$ of energy and materials to produce or process a local resource $L_j$ (that is either an output $L_{out-j}$ or an input $L_{in-j}$ depending on which one is more available or relevant). Hence a specific consumption rate of the operation $j$, $r_j$, is defined:

$$r_j = \frac{E_j}{L_j}, \tag{4.2.2}$$

The choice of the flow $L_j$ is guided in order to have $r_j$ nearly invariant over a large range of $L_j$ for a given process, or at least less variable than the global rate $R_j$.

The modeling of the consumption rate $r_j$ of the subsystem $j$ is independent of the overall system and subsequently can be used in systems different from the first one. The methodology offers thus a modular approach. The information and data to establish the rate can come from sources other than that of the system under study.

The LCA database reports the consumption to manufacture each product $j$ per unit of its mass in g. This is the local rate $r_j$ expressed in $\mathrm{MJ_{EP}{\cdot}g_j^{-1}}$. Hence the conversion into the consumption of operation $j$ at the level of the phone, $R_j$ in $\mathrm{MJ_{EP}{\cdot}phon^{-1}}$, is obtained from the relation $R_j = r_j \cdot w_j^{phon}$. The quantity $w_j^{phon}$ represents the weight of the component $j$. The student had thus to weigh each of the phone components.

For any operation $j$ of a system *syst* the relation between $R_j^{syst}$ and $r_j$ is:

$$R_j^{syst} = r_j \cdot w_j^{syst}, \tag{4.2.3}$$

where the quantity $w_j^{syst}$ corresponds to the amount of $L_j$ required for the whole system to produce a unit of its output:

$$w_j^{syst} = \frac{L_j}{L_{out}}. \tag{4.2.4}$$

The quantity $w_j^{syst}$ provides the link between the local operation and the system. It is also considered the weight of $r_j$ in the overall rate $R^{syst}$ and one of the variables of $R^{syst}$. The rate $r_j$, supposedly less variable than $R_j$ and hence than $w_j^{syst}$, is considered as a parameter, which depends on the choice of the processes for the operation $j$.

### 4.2.2. Critical Analysis of the LCA Database; Choice of the Local Rates

**Sum of and Comparison between the Global Rates $R_j$** About 96% of the total mass of the phone was investigated (total mass versus sum of the masses), and probably more than 96% in terms of $R_j$ as the remaining components are more common material (plastics...). Table 4.2.1 lists by importance the four most energy consuming parts of the phone. The sum of $R_j$ of all weighted components amounts to $R^{phon} = 1\,270$ MJ$_{EP}$·phon$^{-1}$.

**Table 4.2.1. Local and global rates of consumption to manufacture the most energy consuming parts of an IP phone, as well as their mass, from a commercial database**

| Part $j$ | $r_j$ MJ$_{EP}$·$g_j^{-1}$ | $w_j$ in g | $R_j$ MJ$_{EP}$·phon$^{-1}$ | $R_j$ % of $R^{phon}$ |
|---|---|---|---|---|
| liquid crystal display | 14.3 | 62 | 880 | 69% |
| loud speaker display | 19.3 | 7.7 | 150 | 12% |
| plastic shell | 0.135 | 810 | 110 | 8.6% |
| ships | 5.5 | 8.7 | 47 | 3.6% |

By far the most important item is the liquid crystal display LCD. It is a 5 inch (13.5 cm) display with a surface of 88 cm$^2$ and a resolution of 145 x 200 pixels, or 330 pixels per cm$^2$. The liquid crystal used is one of the simplest, the twisted nematic, operating between two states, black and white. The different cells are printed on a simple floppy film sandwiched between two sheets of glass. There is no source of light as it uses the ambient light reflected by one of the sheets acting as a rear mirror. The database offers few information on the origin of $r_{LCD}$; the data was obtained in 2003 from three Asian makers of LCD twisted nematic. There is no information on the resolution or on the existence of a source of light. Interestingly in this database the value of $r_{LCD}$ is similar to the one of a sophisticated color LCD using the technology of Thin Film Transistor TFT and lit by built-in sources. There is no detail on the operations taken into account from the extraction to the transport of the final product, neither on the direct energies and their conversion into primary energies.

The rate $r_{LS}$ of the loud speaker, the second most consuming item, was taken from another commercial database, Ecobilan. The value dates back from 1995.

No other information was available.

The rates $r_j$ of these both items are at least one hundred times as high as the one to manufacture the phone shell, which represents the third largest consuming item because of its mass. The value of the rate for shell is plausible, albeit still large, by comparison with the heating value of its material, light petroleum products, about 50 $MJ_{HHV} \cdot kg^{-1}$.

**Analysis of the Largest Rate $R_j$. Its Local Rate.** Because its consumption represents nearly 70% of the overall phone requirements, the phone display occupies a central place in our analysis.

In fact its mass results mainly from that of the glass sheets. A plausible glass is the common soda-lime glass, which is made out of a so called float furnace using a molten tin bath to form the sheets [47]. The total furnace consumption $r_{glass}$ is about 6.5 $MJ_{HHV} \cdot kg_{glass}^{-1}$ of heavy fuel (from an internal report of St Gobain Recherche, the R&D unit of the glass maker; see also [47]). Taking into account other consumptions such as for the transport and raw material grinding, as well as for the production of the heavy fuel, the overall local rate $r_{glass}$ is about 10 $MJ_{EP} \cdot kg_{glass}^{-1}$. Assuming that the energy consumption of the phone LCD coincides with that of its glass, $R_{LCD}$ would amount to about 0.6 $MJ_{EP} \cdot phon^{-1}$, or at least one thousand times lower than the value deduced from the LCA database.

Either the rate $r_{LCD}$ is still defined with the mass of the whole display, and therefore the energy to produce the film is negligible relative to the energy for glass, or a more correct local rate must be chosen, such as consumption per unit of surface in $MJ.cm_{LCD}^{-2}$ or that par unit of pixel in $MJ.px_{LCD}^{-1}$. In absence of more details from the database, data rom other sources are required to derive their values.

According to a seemingly different source, the monitor of a computer made of a 15 inch color LCD with TFT technology requires 2073 $MJ_{EP}$ [48]. It has a surface of 705 $cm^2$ and about 1024x768 pixels, or 1100 pixels per $cm^2$, with at least 24 bits per pixel. Assuming that the consumption to make a LCD is proportional to its surface, the rate $r_{LCD}$ is 2.9 $MJ_{EP}.cm_{LCD}^{-2}$. Hence the global rate $R_{LCD}$ amounts to about 250 $MJ_{EP} \cdot phon^{-1}$. With the assumptions that the consumption is proportional to the resolution in pixel, the adapted rate $r_{LCD}$ is 2.6 $kJ_{EP} \cdot px_{LCD}^{-1}$. Hence the global rate $R_{LCD}$ would amount to about 75 $MJ_{EP} \cdot phon^{-1}$, ten times as low as the value reported in Table 4.2.1.

The main manufacturer of TFT LCD, Samsung electronics, provides a direct source of data. In its 2009 sustainability report, the company indicates for 2008 the overall consumptions in electricity and natural gas of its factories over the world, mainly in South Korea [49]. They amounted to 11.6 $GWh_e$ of electricity and 1.76 $GWh_{HHV}$ of natural gas. In fact an error of units was made, GWh instead of TWh. The green house emissions in the report allow to correct this mistake, knowing the system of electricity production in South Korea, 170 $TWh_e$ from coal, 80 $TWh_e$ from gas, 130 $TWh_e$ from nuclear fission and 20 $TWh_e$ from hydroelectric generators. Samsung electronics consumes thus 2.5% of the country production, which is at the same time large - difficult to make an error - and small - given their share of the world production of electronic and telecommunication goods.

The company activity is organized into five divisions. The two most consuming activities, 90% of the energy, are the semiconductor (memory chips...) and TFT LCD businesses, requiring consequently 10.3 $TWh_e$ and 1.5 $TWh_{HHV}$ in 2008. The press news of the company give the capacity of its LCD factories in terms of surface of glass panel. For 2008 the capacities were 4.94 $M{\cdot}m^2$ of glass panels in the 7th generation and 3.30 $M{\cdot}m^2$ of glass panels in the 8th generation. Unfortunately the report does not provide the energy break down between the two activities, and the semiconductor capacity in terms of surface of chips is not referenced.

Consequently we can only deduce that $r_{D-LCD}$ is lower than 0.45 $MJ_e.cm^{-2}_{LCD}$ and 0.05 $MJ_{HHV}.cm^{-2}_{LCD}$. The factor of conversion $\beta_e^{EP}$ of electricity into primary energy in South Korea is about $\beta_e^{EP} = 2.5$. Taking a rate for LCD manufacture of about 0.3 $MJ_e.cm^{-2}_{LCD}$, or 0.75 $MJ_{EP}.cm^{-2}_{LCD}$ in primary energy, the global rate $R_{LCD}$ would amount to about 65 $MJ_{EP}{\cdot}phon^{-1}$. Other contributions are negligible such as glass manufacture (after converting its local rate into a rate with the same unit as the rate $r_{D-LCD}$). The final rate is at least 10 times as low as the rate deduced from the LCA database (Table 4.2.1).

Furthermore, the display of Samsung electronics is based on techniques of integrated circuit whereas the IP phone screen is made of a printed sheet, of which processes are simpler and less energy consuming.

As we were not able to pursue the study for lack of more accurate data, a large uncertainty remains. But a conclusion stands firmly: large and often unsuspected errors - from a factor 10 to 1 000 - are possible and even probable in absence of sufficient details, and/or independent methods and sources to obtain a value for the same quantities.

## 4.3. The Agro-Ethanol Industry. Analysis of the System

A more complete application of the methodology is provided with the example of the agro-ethanol industry, more specifically the consumptions at the farm where crops for ethanol are produced [12]. The whole study aims to compare efficiency between four different crops for the industry input $L_{in}$, two from sugar plants - sugar cane and beet - and two from cereals - corn and wheat -, in three countries [50]. The section 4.3.1 presents rapidly the global context of this industry as well as the various constraints - economic and environmental - it must meet.

The methodology first fixes the boundary of the system (as for a LCA), here from farm to ethanol distribution. For a more complete study from the field to the vehicle wheels see the article [50] and references therein. The global rate $R$ of the agro-ethanol industry is then defined as the external energy required per unit of LHV of ethanol produced in $\mathrm{J_{EP} \cdot J_{etoh}^{-1}}$. Other choices for the rate of this system are presented in the section 2.2 of the chapter 2.

Fig. 4.3.1 describes the whole system in terms of its main operations and the energy and material flows at its boundary and between its operations. Fig. 4.3.1 also shows the flows of energy in external system auxiliaries to supply the direct requirements $E_D$ of the main operations.

Fig. 4.3.1 shows two types of decomposition, which are dealt with successively in the sections 4.3.2 and 4.3.3:

- a decomposition in series of the main system as in the phone study,
- a transverse decomposition or separation between the main operations of the system and the auxiliary ones.

The aim of both decompositions is to direct the study towards simpler and independent systems as raw data to understand the overall system consumption are encountered at this level. As each consumption $E_{D-j}$ of the operations of the main system has its specific auxiliary system - electricity or fuel productions, fertilizer manufacture... -, the decomposition in series is performed first.

The example also shows how the methodology provides the means to save analysis work. The study insists again on the risk of important errors without

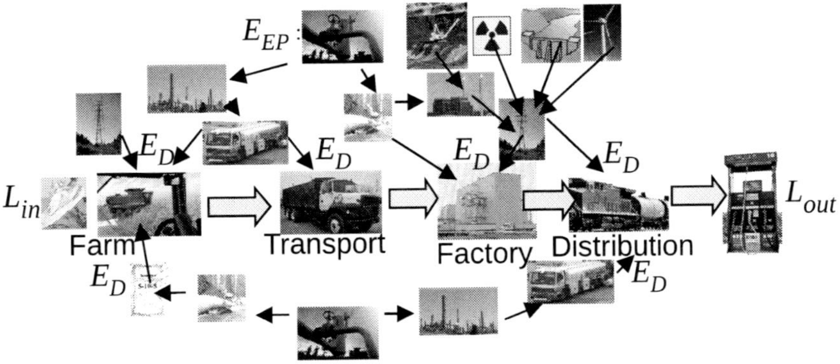


Figure 4.3.1. Main operations of the agro-ethanol industry as well as its external auxiliaries, chiefly petroleum, gas and electricity productions. The input flow $L_{in}$ represents the agriculture harvests, while the output $L_{out}$ is the LHV of the corresponding production of ethanol. The black arrows represent the flows in external operations from primary energy $E_{EP}$ to direct requirements $E_D$ of main operations.

thorough analysis. It also gives the tools to estimate the uncertainty on the rate $R$ from the uncertainty on the raw data.

### 4.3.1. A Short Presentation of the Global Context and Industrial Constraints

In 2009 the world production and consumption of agrofuels (from agriculture crops such as cereals, sugar plants and oleaginous) represented 52 $\text{Mtoe}_{\text{LHV}}$ in heat values. By comparison the production of crude oil amounted to 3900 $\text{Mtoe}_{\text{LHV}}$ (from Figure 3.2.1 in the chapter 3), of which more than half is transformed to fuels for transport. Among agrofuels, agro-ethanol production reached nearly 38 $\text{Mtoe}_{\text{LHV}}$. The USA and Brazil are the two main producers, with 20 and 13 $\text{Mtoe}_{\text{LHV}}$ respectively in 2009. Corn is mostly used in the USA, while Brazilian producers of ethanol process sugar canes. By comparison ethanol production in France from wheat and sugar beet reached 0.64 $\text{Mtoe}_{\text{LHV}}$ in 2009.

Although small by comparison with the fossil oil production, the agro-ethanol production has an impact on the worldwide cereal supplies, and subsequently on the trade prices of these commodities. The harvests, and consumptions, of the three major cereals in the world for 2009 amounted to about 820 Mt for corn, 690 Mt for wheat and 685 Mt for rice (according to the statistics of the Food and Agriculture Organization). The USA is the major producer and exporter of corn and wheat. According to the US department of Agriculture, 20 $\mathrm{Mtoe_{LHV}}$ of ethanol required around 110 $\mathrm{Mt_A}$ of corn grains in 2009, which amounted to a third of the USA corn harvest. The American industry produces from the corn residues at the factory an animal feed of which nutrient and caloric values are equivalent to that of corn grains of the same mass for cattle (ruminants are able to digest fibers either from residues or grains). In 2009 the residue quantity represented 35 Mt and roughly saved a similar amount in corn grains. An additional export of 75 Mt of corn, equivalent to 12 days of global consumption of cereals, would have been significant.

In terms of surface, with an agriculture yield $Y_A$ close to 10 $\mathrm{t_A \cdot (ha \cdot y)^{-1}}$, the 2009 agro-ethanol production in the USA required about 11 Mha of good crop land. Grain corn necessitates a specific climate during its cycle from around April to October, humid, warm and dry at the end of the cycle, without excess. These conditions are mainly encountered in the north of the middle east of the USA around the Great Lakes, the so-called Corn Belt. Corn is grown elsewhere like in the middle west plains - Kansas, Nebraska... -, or in the west coast but usually requires much more irrigation, which has an impact not only in terms of water resource but also in terms of energy consumption of the pumps [12, 50]. These restrictions have an influence on the processes used and limit the resource availability, here the corn planted surface.

The different operations of the industry under study obey, or tend to obey, some constrains. For economic reason - labor and surface productivities - the farm operations correspond to intensive and mechanized agriculture that requires synthetic fertilizers and heavy equipment. Some plant residues (leaves, straw or corn stover) must be left to replenish the soil with organic matter or humus, at least 6 $\mathrm{t_{DM} \cdot (ha \cdot y)^{-1}}$ in dry basis. This action limits the residues available for industrial use such as fuels to power ethanol production at the factory. On the other hand, it avoids large run off of nutrients, and, hence, contamination and the excessive requirements of fertilizers. Similarly some lime must be added depending on the pH of the soil. The phase out of trash burning in sugar cane fields of Brazil is having benefices both in terms of soil preservation and air contamination. It requires more mechanized operations, but allows residues and nutriments surplus.

At the factory stage all the effluents must be treated such as the stillage resulting from the ethanol distillation. If not used as feed or fuel, its demand in oxygen must be at least satisfied. When the concentrates of stillage are dried, the resulting emission of volatile organic components must be neutralized [51]. Heavy investments may limit the use of efficient integrated processes, as it is the case in Brazil at the difference of European and American factories.

### 4.3.2. Levels of Decomposition in Series

In this subsection, we examine the depth of the decomposition of the system and the flexible role of the local rate $r_j$ of an operation $j$, which can become the global rate $R^j$ of the operation considered as an independent and complex system to be studied.

The analysis is carried out with the direct rates $R_{D-j}$ and $r_{D-j}$. For simplification in the notations the subscript $D$ will be omitted. On the other hand, an exponent on the symbol of the quantities, like in $R_j^{syst}$, indicates which system or level of decomposition is studied. Exponents or subscripts may appear cumbersome, however, they contain the necessary information to situate the operation in the overall system.

#### At the Level of the Industry

The agro-ethanol industry $etoh$ is split into four main operations, the farm, the transport of the harvest to the factory, the plant processing into ethanol in the factory, and the ethanol distribution down to the gasoline outlets:

$$R^{etoh} = R_{Fm}^{etoh} + R_{Tra}^{etoh} + R_{Fct}^{etoh} + R_{Dis}^{etoh}.$$

For each main operation further decompositions is achieved. At the farm the consumptions result from the use of a tractor, the application of N fertilizer, the irrigation...:

$$R^{etoh} = R_{TrcFm}^{etoh} + R_{NFm}^{etoh} + R_{IrrFm}^{etoh} + \dots + \dots$$

And even further, with the diesel consumption of the tractor and the energy cost of its fabrication:

$$R^{etoh} = R_{dislTrcFm}^{etoh} + R_{fabTrcFm}^{etoh} + \dots$$

Faced with theses different levels of decomposition, one of the two following strategies can be chosen:

- to work directly at the finest level of decomposition and find the local rates of the elemental operations defined at this level,
- or to work progressively by introducing first the local rates at an intermediate level, and consequently for large operations.

This choice depends on the desired depth of decomposition, which is guided by two considerations:

- to work at the level of an elemental process in order to derive an adapted $r_j$, which is nearly independent of any variable and therefore at lower risk of error and uncertainty,
- to obtain the data corresponding to this level of decomposition from different sources. The deeper is the decomposition, the larger is the number of operations to deal with, the more specialized is the information and subsequently the longer is the time required to gather and analyze the data.

A compromise must be found. The strenuous work of gathering and analysis of data for an operation $j$ will depend on the impact of its rate $R_j$ on $R$ measured by the ratio $R_j/R$, and so on the precision required for $R_j$. The larger is the share of $R_j$, the more thorough is the analysis and the deeper is the decomposition.

Faced with this compromise between time and accuracy, the second strategy appears more adequate. In practice an iterative work is done. A first set of data at a coarse level of decomposition allows a rough determination of the rates $R_j$. Largest rates are further worked out with thinner decomposition and more accurate data. The values of the lower rates are considered acceptable even with large bar of errors (factor of more than 2). The example of the tractor at the farm shows the application of this strategy.

### At the Level of the Farm

The farm produces the plant used by the industry. Hence the local rate $r_A$ of the consumptions at the farm is defined per unit of the harvest in $\mathrm{t_A}$.

The conversion of the local rate into the global rate $R^{etoh}_{Fm}$ requires the quantity $w^{etoh}_A$ according to eq. 4.2.3. Its unit is deduced mechanically from the units of the rates $r_A$ and $R^{etoh}_{Fm}$: in $\mathrm{t_A \cdot J^{-1}_{etoh}}$, *i.e.* the amount of plant required to produce 1 J of ethanol. The inverse of $w^{etoh}_A$, in $\mathrm{J_{etoh} \cdot t^{-1}_A}$, is expressed as:

$$(w^{etoh}_A)^{-1} = x_{etoh} \cdot LHVm,$$

where $x_{etoh}$ is the ethanol yield at the factory, in $\mathrm{kg_{etoh}{\cdot}t_A^{-1}}$, and LHVm is the LHV of ethanol per unit of mass, exactly 26.8 $\mathrm{MJ_{etoh}{\cdot}kg_{etoh}^{-1}}$ (LHVm of a unique molecule).

The weight $w_A^{etoh}$ depends thus on variables fixed by the operations after the farm, chiefly the factory. This is a general rule when working with global rates based on the output: because the local rate $r_j$ is specific to the operation $j$ whereas the rate $R_j$ is linked to the whole system under study and its output, the terms $w_j$ of upstream operations depend on quantities of the downstream operations representing their losses or factors of conversion.

The ethanol yield $x_{etoh}$ is broken down into two important physical factors, the content of sugar or starch in the plant or grain, $x_S$, and the yield in mass to convert it into ethanol at the factory. For a fixed mass of sugar/starch the ethanol yield is limited by the stoichiometry of the overall reaction of the glucose transformation into ethanol. In the case of cereals, the hydrolysis reaction of starch into sugar with the addition of water molecules must be also taken into account. The theoretical ethanol yield, the Gay-Lussac yield, is 0.511 and 0.568 kg of ethanol per kg of sugar and starch respectively. In practice other reactions decrease ethanol yield, which is measured by the factor $\alpha_{etoh}$, the fraction of sucrose or starch converted into ethanol (Table 4.3.2). Even in very controlled reactors unavoidable reactions (sucrose conversion into glycerol, about 3% of the sucrose, yeast nutrients, about 1%, fusel oil...) limit the ethanol yield to the so-called Pasteur yield, which represents 94.7% of the Gay-Lussac yield or $\alpha_{etoh} = 0.947$. Hence:

$$(w_A^{etoh})^{-1} = (0.511 \text{ or } 0.568) \cdot \alpha_{etoh} \cdot x_S \cdot LHVm. \qquad (4.3.1)$$

Table 4.3.2 reports for each culture the physical and technical quantities in eq. 4.3.1 specific to the plants and the industry processes, as well as other factors introduced further down - harvest yield $Y_A$, moisture of the harvested plant $H$, nitrogen content $x_N$. All of them represent potential variables for the rate of the system $R$.

To go further in the decomposition, we consider now the farm as the main system, independent of the agro-ethanol industry. The farm rate $r_A$ becomes the global rate $R^A$ in $\mathrm{J}_D{\cdot}\mathrm{t_A^{-1}}$. The rate $R^A$ is the sum of the consumption rates for the use of tractor, the application of N fertilizer, the irrigation...:

$$R^A = R^A_{TrcFm} + R^A_{NFm} + R^A_{IrrFm} + \cdots$$

**Table 4.3.2. Plant and factory variables for $w_A^{etoh}$, $w_N^A$ and $w_{pump}$**

| | $Y_A$[a] | $H$[b] | $x_N$[c] | $Y_N$[d]. | $x_S$[e] | $\alpha_{etoh}$[f] | $x_{etoh}$[g] |
|---|---|---|---|---|---|---|---|
| Cane | 72.7 | 70 | 1.37 | 92 | 143 | 0.86 | 63 |
| Beet | 66.2 | 77 | 1.60 | 100 | 180 | 0.86 | 79 |
| Corn | 8.7 | 15 | 15.0 | 150 | 620 | 0.92 | 320 |
| Wheat | 9.0 | 15 | 18.5 | 185 | 570 | 0.91 | 295 |

[a]harvest yields averaged over a culture cycle and a region expressed in $t_A \cdot (ha \cdot y)^{-1}$, according to the references in the article [12].

[b]water content of harvested plants in %mass, as indicated by the references or by national agriculture statistics. For cereals $H$ also corresponds to the minimal moisture for a long time storage. It can differ from the actual content at the field.

[c]the fraction of nitrogen N in the harvested plant in $kg_N \cdot t_A^{-1}$.

[d]the reported application of N fertilizer in $kg_N \cdot (ha \cdot y)^{-1}$

[e]the average fraction in $kg_S \cdot t_A^{-1}$ of sugar or starch in harvested plants as received at the factory (with the moisture content $H$ of the table).

[f]the fraction of sucrose or starch converted into ethanol. The values represent the recent and probably highest performances.

[g]the ethanol yield at a modern factory in $kg_{etoh} \cdot t_A^{-1}$.

Let's examine each of them, and let's apply the same method as for the farm if necessary.

**At the Level of the Tractor** The mechanized operations requires some diesel, representing their operational consumption, as well as a tractor and other equipment, representing the investment consumption. Both consumptions is studied together independently of other farm requirements.

For a same type of operations with a tractor, the consumption is more proportional to the distance traveled, and so the planted surface, than to the mass of the harvest. Hence the local rate $r_{dislTrcFm}$ for diesel consumption is expressed in $J_D$ or liter of diesel per ha and per year (year corresponds often to the period of a culture cycle in a temperate climate). The rates $r_{dislTrcFm}$ for the cultures under study are in the order of 100 liter of diesel per ha·y (from 66 for corn culture in the USA to 180 for sugar beet in France). They present a lower dispersion than working with a rate in liter per $t_A$ [12]. The differences observed between cultures or even between various area of a same culture result from variables other than the surface such as the number of operations during the culture cycle and the consumption rate of each one (the plowing or tillage is the most intensive one). Some sources give sufficient details to resolve these discrepancies. However, the importance of the mechanized operations in the agro-ethanol industry expressed with $R^{etoh}_{dislTrcFm}$ does not deserve this level of scrutiny.

The conversion of the rate $r_{dislTrcFm}$ into the rate $R^{A}_{dislTrcFm}$ at the farm level requires a weight $w^{A}_{ha}$ in $\mathrm{ha{\cdot}y{\cdot}t_A^{-1}}$ according to eq. 4.2.3. It represents the surface necessary each year to produce 1 tonne of harvest, *i.e* the inverse of the yield $Y_A$. The quantity $w^{A}_{ha}$ is used for other consumptions that are also proportional to the cultivated area, such as the energy for irrigation - see below -, seeds and agrochemical applications.

A direct conversion of the local rate $r_{dislTrcFm}$ into the global rate $R^{etoh}_{dislTrcFm}$ necessitates the weight $w^{etoh}_{ha}$ in $\mathrm{ha{\cdot}y{\cdot}J_{etoh}^{-1}}$ or the area required each year to produce 1 joule of ethanol. The relation with other weights is:

$$w^{etoh}_{ha} = Y_A^{-1} \cdot w^{etoh}_{A} = w^{A}_{ha} \cdot w^{etoh}_{A}.$$

The result at the system level is, as expected, identical with or without the intermediary level of the farm.

The investment consumptions correspond to the energy required to manufacture the equipment for the operation $j$. Its natural or local rate $r_{inv-j}$ is expressed in $\mathrm{J_{inv}.equipj^{-1}}$ like in the case of the IP phone seen in the previous section. It differs from the local rates $r_{op-j}$ of the operational consumptions, as these latter are readily related to the output $L_j$ of the operation $j$ and are expressed in $\mathrm{J_{op}{\cdot}L_j^{-1}}$.

To draw a comparison between the two types of consumption, the investment rate $r_{inv-j}$ is converted into the same units as the operational rate $r_{op-j}$. It is a problem equivalent to the conversion, according to eq. 4.2.3, of the local rate $r_{inv-j}$ into the global rate $R^{j}_{inv-j}$ for the independent system $j$, of which units are fixed by those of $r_{op-j}$. The factor of conversion $w^{j}_{inv-j}$ represents the number of equipment per unit of the operation output $L_j$. The inverse has a more straightforward meaning: the amount $L_j$ produced during the equipment lifetime.

The quantity $\left(w^{TrcFm}_{fab}\right)^{-1}$ to convert the local rate $r_{fab}$ for manufacture into the rate $R^{TrcFm}_{fab}$ at the level of the mechanized operations at the farm, $TrcFm$, represents the number of hectares cultivated by the tractor during its lifetime. A tractor manufacturer claims a lifetime of 22 000 hours for one of its tractors operating in sugar cane fields [11], probably a typical one. In average we assume a lifetime of 18 000 hours. Trials in wheat fields indicate a period of use of the tractor of about 5 h·(ha·y)$^{-1}$. Consequently the weight $(w_{fab}^{TrcFm})^{-1}$ is around 3 600 (ha·y)·tractor$^{-1}$, which is equivalent for the tractor to operate for 18 years in a farm of 200 ha.

The value of $w_{fab}^{TrcFm}$ is a rough estimate with an uncertainty at least of 30%. However, the section 4.3.4 shows that it is sufficient in the case of the agriculture operations. In other industries such as solar, wind or even nuclear energies the equipment can represent the major consumptions. As the investment contribution is usually more difficult to estimate, we have to expect a larger uncertainty on the final results than for the operational consumptions, or a long work of data gathering and analysis.

**At the Level of the N Fertilzer** The application $Y_N$ of N fertilizer depends on the fraction of N in the harvested plant $x_N$ in $\mathrm{kg_N \cdot t_A^{-1}}$ in form of proteins, adenosine triphosphate ATP and deoxyribonucleic acid DNA (see Table 4.3.2 for the data). A choice for the rate $r_N$ is then the mass of fertilizer required in $\mathrm{kg_N}$ per kg of N in the plant. Agronomic studies show that the balance of the element N from inputs and outputs - also including the change of natural N storage - is difficult to establish precisely due to the various and not well-known inputs (from natural phenomena or from previous cultures) and losses (runoff or gaseous emissions) [12]. The losses depend in part on the agriculture practices (cover to limit runoff). According to trials in England with a continuous high yield wheat culture (limit the contribution of other N inputs) and with averaged conditions, the losses between the N application and the nitrogen content in the plant amount to 20% of the input, or a rate $r_{D-N} = 1.25$ $\mathrm{kg_N}$ per kg of N in the plant. It more or less agrees with the data of other cultures.

The weight $w_N^A$ to convert into the rate at the farm level is simply the N content $x_N$.

The weight $w_N^{etoh}$ to convert into the rate at the industry level is expressed in $\mathrm{kg_N}$ of N in the plant per $\mathrm{J_{etoh}}$, which is related to $w_A^{etoh}$:

$$w_N^{etoh} = w_N^A \cdot w_A^{etoh} = x_N \cdot w_A^{etoh}.$$

As a side note the element N is absent from the molecule of ethanol, and consequently useless for its production, except the minimum requirements for the existence of the plant (DNA, ATP...). The content $x_N$ is also an important quantity to estimate the amount of residues to be treated at the factory, and hence the cost of this treatment. All of these observations give incentives to reduce the content $x_N$ such as the use of sugar plants instead of cereals.

**At the Level of the Irrigation** In Nebraska close to 60% of the corn planted area are irrigated with water pumped from an underground aquifer. From data of the US irrigation surveys state by state, the average head pressure from the pump in Nebraska is about 1.0 $\mathrm{MJ}_m \cdot \mathrm{m}_{\mathrm{wp}}^{-3}$ of pumped water [12]. The aquifer is unconfined with nearly no artesian pressure and therefore at the atmospheric pressure. The water is lifted from an average depth of 25 m and pressurized at 0.39 MPa at the well head.

From fluid mechanics, the direct work required from the pump is proportional to the volume of the pumped water and to the head pressure, or a theoretical local rate $r_{m-pump} = 1.0\ \mathrm{MJ_m} \cdot (\mathrm{m}_{\mathrm{wp}}^3 \cdot \mathrm{MPa_{head}})^{-1}$. The head provides for the lift of water, its pressurization and the friction losses resulting from the flow. Without the friction losses, the theoretical mechanical energy required in Nebraska fields would amount to 0.54 $\mathrm{MJ_m} \cdot \mathrm{m}_{\mathrm{wp}}^{-3}$. In fact, the pumps must apply a suction to increase the water throughput, enhancing the friction of the flow within the porous rock and making it an important contribution to the head pressure [12]. This contribution, approximately proportional to the throughput, is taken into account with a factor proportional to the pressure fall due to the water lifting and pressurization. Hence the actual mechanical rate $r_{m-pump}$ is 1.9 $\mathrm{MJ}_m \cdot (\mathrm{m}_{\mathrm{wp}}^3 \cdot \mathrm{MPa_{head}})^{-1}$ (where the head pressure includes only lifting and pressurization), leading to the integrated rate $r_{m-pump} = 1.0\ \mathrm{MJ}_m \cdot \mathrm{m}_{\mathrm{wp}}^{-3}$ for the average well in Nebraska.

From the US irrigation surveys the volume $w_{pump}^{ha}$ of the pumped water per unit of irrigated area in Nebraska for 2003 represented 3 700 $\mathrm{m}_{\mathrm{wp}}^3 \cdot (\mathrm{ha} \cdot \mathrm{y})^{-1}$, equivalent to 370 mm of rainfall a year (when natural one is about 700 mm). The weight for pumped irrigation at the farm is $w_{pump}^A = w_{pump}^{ha} \cdot w_{ha}^A$, with $w_{ha}^A$ from section 4.3.2.

As seen above the rate $r_{m-pump}$ in $\mathrm{MJ}_m \cdot \mathrm{m}_{\mathrm{wp}}^{-3}$ depends on the mechanical characteristics of the water source (depth, pressure, proximity from fields...). The weight $w_{pump}^{ha}$ varies according to the adequacy of the cultivated plant to the climate in the area under study. Consequently the rates $R_{IrrFm}^A$ and $R_{IrrFm}^{etoh}$ depends on each situation and the value for Nebraska cannot be generalized.

### 4.3.3. Conversion into Primary Energy

The conversion into primary energy refers to the production by auxiliary systems of the material and energy requirements of the main operations. They comprise the operations of extraction from natural resources and their transformation. Their consumptions include the content of the feedstock as well the energy losses in auxiliary operations.

As this conversion is independent of the conversion from the rate $r_{D-j}$ into the rate $R_{D-j}$ of direct requirements $E_{D-j}$, it can be carried out before this step at the level of the local operation $j$, from $r_{D-j}$ to $r_{EP-j}$.

The decomposition between the main system and the auxiliary systems is different from the decomposition used in the section 4.3.2 as the former leads to factors $\beta_{aux-j}$ which form products with the direct rates.

The modular approach again applies for each of the basic auxiliary operations $aux-j$, of which output is the flow $E_{D-j}$. The factor $\beta_{aux-j}$ is determined from the rate $R^{aux-j}$ of the auxiliary studied as an independent system using eq. 2.1.3 (see the section 2.1.3 of the chapter 2).

In theory the conversion is carried out for each direct requirement $E_{D-j}$ of each main operation $j$. Practically a large part of the direct requirements is produced by a limited numbers of these auxiliary systems, such as electricity production - depending on the primary energy used -, petroleum industry, gas extraction and processing, manufactures to make chemicals, steel, glass, cement... This limited number allows to constitute a database of basic operations and their processes. The last operations to deliver the direct requirements of the main operations may be located in the same area and be dependent on them.

In the cases of the diesel consumptions for mechanized operations and the N fertilizer application, the auxiliary systems are the diesel production from crude oil and the production of N fertilizer from natural gas *via* the ammonia factory, respectively. Diesel and ammonia can be produced from other feedstock by various routes of processes, hydrocarbons synthesis from natural gas, biomass or vegetable oil for diesel, and from coal, heavy fuel or electricity for ammonia. However, these systems of production are so far marginal for technical and economical reasons.

**At the Level of the Tractor** The consumption rate to produce diesel from crude oil is approximated to $\beta_{dsl} = 45\ \mathrm{MJ_{EP}} \cdot \mathrm{l}_{\mathrm{diesel}}^{-1}$, including the volumetric heat value of diesel - 36 $\mathrm{MJ_{LHV}} \cdot \mathrm{l}$ - and the fuels consumed for the operations - nearly all from internal sources - [20]. The factor $\beta_{dsl}$ of this energy system is thus $\beta_{dsl} = 1.25$. There is not a unique value for the petroleum industry, depending mainly on processes used at the refinery to fulfill demand and regulations, and on the maturity or specificity of each oil field (see the end of the section 3.1 in chapter 3).

The primary energy requirement to manufacture the tractor is calculated in the next section about errors and uncertainty.

**At the Level of the N Fertilzer** The coefficient of conversion $\beta_N$ for N fertilizer takes into account the consumptions to treat and transport the gas to ammonia factories, as well as the final form in which the fertilizer is delivered (straight ammonia, ammonium-nitrate, urea...) [12]. The total value thus differs but is around 45 $\mathrm{MJ_{EP} \cdot kg_N^{-1}}$, based on the best available technology in modern factories [9, 12].

**At the Level of the Irrigation** The engines which drive the pumps require either diesel, electricity, or gas. Thus the coefficient $\beta_{xpump}$ comprises the efficiency of the processing chain for each final energy, as well as their share of the total mechanical rate $r_{m-pump}$ and the efficiencies of engines, pump and mechanical transmission. The efficiency of each leg $k$ of the chain is taken into account by its factor $\beta_k = E_{ink}/E_{outk}$, the inverse of its energy yield $Y_k$. The coefficient $\beta_{xpump}$ is the product of these factors.

The factor $\beta_{EP-Eng}$, from primary energy to pump inlet, corresponds to a weighted average of the consumptions of the chains of production of the three final energies. Electricity is itself a weighted average of the different primary energy used for its production. Technical and statistical data must be gathered to establish the factors. For instance the diesel production from oil has a factor $\beta_{dsl} = 1.25$,the yield of a typical diesel engine can reach 42.5% - $\beta_{dslEng} = 1.81$ -, and its share of the engine output is 38% [11]. The overall factor $\beta_{EP-Eng}$ of the first operations of the chain is 3.3. The transmission efficiency between the engine outlet and the pump is 94%, or $\beta_{tran} = 1.06$, and the conversion into head pressure is about 75%, or $\beta_{pump-m} = 1.33$. The product $\beta_{xpump}$ amounts thus to 4.6 $\mathrm{J_{EP} \cdot J_m^{-1}}$.

### 4.3.4. Errors and Uncertainty

**Types of Errors** A first type of errors comes from the raw data themselves. The access to data through original sources is important to obtain more details and to have lower risk of errors than with indirect sources. The genuinely different sources should be used and their data compared to track errors.

For instance two publications about the ethanol industry in Brazil present a discrepancy for the heavy-duty truck load of sugar cane [20]. Consideration on the apparent density of the sugar cane and the level of precision provided by one source allow to retain its data.

A similar discrepancy is observed for the fraction of cropland irrigated in Nebraska [12]. In 1998 it amounted to 60% according to the state surveys, or to 48% according to the federal ones. However, the data from both sources for other years show a stable value at around 60%. A reporting error has probably occurred.

In another case, the total diesel consumption for sugar beet culture varies by more than 50% from one publication to another [12]. One of them reports the data of another publication taken in a different country, and consequently seems less reliable.

On the other hand, the difference may result from a key variable. However, in the case of sugar beet culture available information was insufficient to identify it.

Another illustration of the importance to investigate differences is the state averaged consumptions of corn growers in 9 rural states of the USA, as reported in the 5 year national agriculture census until 2001 [11]. They show that consumption in the Nebraska state was about three times as high as those in the 8 other states. Its consumption also varied from one census to another, depending on the meteorological conditions during the census year. A thorough data gathering and analysis show that pumped irrigation is widespread in Nebraska whereas negligible in other states. Thus pumped irrigation is a key variable of consumptions at the farm. This example shows the danger of data corresponding to a marginal producer, or to an average over a large number of producing units. The average data may comprise too many different operations or processes, and hide the influence of variables. The average can even have no meaning at all. Better data are the energy and material flows for a large group of factories using the same processes like in the case of Samsung Electronics for the study of liquid crystal display manufacturing. Details on the data origin can dispel such ambiguities.

Another source of information, which also provides a means to check data from industrial or public sources, is the established relations and tabulated values in sciences and industry. They have demonstrated their validity over a large amount of experiences, industrial processes... and permit to design a chain of operations or extrapolate the inputs and outputs of a pilot at larger scales. Such information are the various laws of conservation like that of mass and elements, the thermodynamics laws (among which the energy conservation), the fluids mechanics equations (deduced from the momentum conservation), the friction laws, the empirical Bond relation to deduce the consumptions of grinding...

Another type of errors is the incorrect interpretation of the original data or their lack of precision. It introduces uncertainties lower than 50% of the reported value, but cumulated may amount to a factor of 2. The analysis of the agro-ethanol industry offers a number of them:

- the cycle of sugar cane culture lasts in average 6 years with 5 harvests. How to compare it with a cycle of one year?
- there is a difference between the high and low heat values for hydrocarbon fuels (10% for methane, 17% for $H_2$),
- the humidity of the harvested cereals in the field often differs from the normalized humidity used to report $Y_A$ in the national statistics, or from the one in the grains received at the factory,
- the ethanol output at an American factory may include a denaturant added at the end of processes (when there is no indication it is often included; the yield $x_{etoh}$ allows to discriminate),
- the N fertilizer applications as reported by farmers ($Y_N$ in table 4.3.2) can lead to a N content in plants different from the actual content $x_N$ deduced from a database of plant constituents, even by taking into account the losses of N owing to the rate $r_N$,
- ...

**Estimate and propagation of uncertainties** After discarding possible errors by the previous analysis, some uncertainties on the raw data remain. At the difference of the significant influence of key variables, the uncertainty represents unresolved fluctuations between data due to the weak dependence on multiple parameters not taken into account. It also models the consequences of a lack of detailed information. By reasonable assumptions on the apparatus or on the method of measuring and reporting data, or by other means such as deriving the same information from different sources, the uncertainty on the raw data are evaluated. Their contribution to the accuracy of $r_j$ and $w_j$, and hence on $R_j$, are estimated thanks to the error propagation theory based on independent stochastic variables. The standard deviation $\delta x$ is used to express the data uncertainty on quantity $x$ (either the raw data or the final rate). As a matter of fact, due to the small amount of data to calculate it, the standard deviation $\delta x$ for one variable is often a mere estimate.

The two main relations between uncertainties $\delta x$ used are deduced from the relations between their associated variables $x$. For the rate $R$ as the sum of $R_j$, the relation between their uncertainties is:

$$(\delta R)^2 = \sum_j (\delta R_j)^2 \tag{4.3.2}$$

For the rate $R_j$ as the product of the local rate $r_j$, the factor $\beta_j$ and the weight $w_j$, the relation between their deviations is:

$$\left(\frac{\delta R_j}{R_j}\right)^2 = \left(\frac{\delta \beta_j}{\beta_j}\right)^2 + \left(\frac{\delta r_j}{r_j}\right)^2 + \left(\frac{\delta w_j}{w_j}\right)^2 \tag{4.3.3}$$

If one of the relative deviations $\delta x/x$ in the right hand of eq. 4.3.3, or one of the absolute deviations $\delta x$ in eq. 4.3.2, is larger than other deviations by a factor of at least two, the resulting deviation at left hand is approximated to the dominant relative, or absolute, deviation.

**Example of the Tractor at the Farm (see the section 4.3.2)** The local rate $r_{fabTrc}$ to manufacture a tractor is estimated from the masses of its main constituents. A typical modern tractor is made of 6 tonnes of steel and about 0,5 t of tires (other materials are neglected). Taking into account the factors of conversion from the auxiliary systems, roughly $\beta_{aux}$ of 15 $\mathrm{GJ_{EP}{\cdot}t_{steel}^{-1}}$ and 50 $\mathrm{GJ_{EP}{\cdot}t_{tire}^{-1}}$ for recycled steel and petrochemical products respectively, the local rate $r_{fabTrc}$ in primary energy amounts to about 100 $\mathrm{GJ_{EP}{\cdot}tractor^{-1}}$. We assume an error of 50 $\mathrm{GJ_{EP}{\cdot}tractor^{-1}}$, or 50% for $\delta r_{fabTrc}/r_{fabTrc}$.

Hence the rate $R_{fab}^{TrcFm}$ has an estimated value of 30 $\mathrm{MJ}_{EP}.(\mathrm{ha{\cdot}y})^{-1}$, taking into account the weight $w_{fab}^{TrcFm}$ previously calculated. If the relative uncertainty on $w_{fab}^{TrcFm}$ is lower than that on $r_{fabTrc}$, let's say 30%, the cumulative relative error on the rate $R_{fab}^{TrcFm}$, $\delta R_{fab}^{TrcFm}/R_{fab}^{TrcFm}$, is also about 50% using eq. 4.3.3.

The consumption rate of diesel $R_{disl}^{TrcFm}$ is around 4 500 $MJ_{EP}\cdot(ha\cdot y)^{-1}$, far larger than the rate $R_{fab}^{TrcFm}$ for investment. For a specific culture its uncertainty $\delta R_{disl}^{TrcFm}/R_{disl}^{TrcFm}$ is lower than 10%. The total rate $R^{TrcFm}$ for the mechanized operations, the sum $R_{disl}^{TrcFm} + R_{fab}^{TrcFm}$, is thus approximated to the operational rate $R_{disl}^{TrcFm}$. From eq. 4.3.2 and the values of rates and deviations, the uncertainty on the overall rate $R^{TrcFm}$ is given by the uncertainty on the operational rate $R_{dsl}^{TrcFm}$, less than 10% in relative terms. A rough estimate of the rate for a small consumption, along with its deviation, is thus sufficient.

## Summary of the Methodology

Even for a well-defined system, owing to its inputs and outputs, and a fixed choice for its efficiency indicator or rate of consumption $R$ (see the potential definitions in chapter 2), a unique value for the rate $R$ is unlikely. Each operation $j$ of the system is achieved by a range of processes with different consumption rates, and for a fixed process the contribution of the operation to the rate $R$ depends on the physical or technical variables of the system. Moreover, the raw data can present a large uncertainty.

A methodology is proposed to identify these influences, which are a source of large errors if undocumented, and hence to built a relation between the efficiency of the system and its physical variables and parameters. The rate becomes thus a tool not only to compare the efficiency of various systems with same output, but also to predict future evolutions.

The methodology relies on the decomposition of the system towards elementary operations at which level specialized and more accurate information is retrieved. Each component $j$ from this decomposition presents its own efficiency measured by a local rate $r_j$ based only on the operation flows and the retrieved information. From the rate $r_j$ is determined the contribution $R_j$ of the operation $j$ to the system rate $R$. The coefficient $w_j$ between both rates of the operation $j$ represents one of the variables of the system.

The methodology also deals with the propagation of uncertainties on raw data to the final results thanks to the relations established between the rates. It allows to optimize the time of analysis devoted to determine the local rate and the variables for each operation $j$.

# Chapter 5

# Energy Systems. Self-reliant Systems

In the section 2.1 of the chapter 2 we have given a practical definition of the efficiency of any industrial system with its energy consumptions per unit of its output (or input if better suited), the rate $R$ (eq. 2.1.2 according to Fig. 2.1.1).

However, this definition is ambiguous for an energy system such as the petroleum or agro-ethanol industries, as already discussed in the chapter 2. The main input, crude oil or agriculture harvest, possesses a heat value and part of it can be dissipated to fuel the system processes. Depending on its exact definition, for instance when taking into account only the consumptions of external energy, the rate cannot capture the dissipated part.

The chapter 2 has introduced another rate, the dissipation rate $R_{diss}^{in}$ (eq. 2.2.1), better suited to energy systems thanks to the use of the energy conservation. By definition its upper value is limited to one. The closer to one the rate $R_{diss}^{in}$ is, the more inefficient the system is. However, the system for which the rate is defined also includes auxiliary operations like electricity production, which supply the external requirements of the main operations of the system. Consequently the efficiency as measured by the rate represents an overall efficiency and not that of the main operations.

The chapter 2 also mentioned the existence of self-reliant energy systems for which the resource they exploit provide both the feedstock and the fuels of their direct processes. This chapter deals with the efficiency of some self-reliant energy systems such as steam engines and the ancient exploitation of coal in underground mines. The latter operates steam engines to provide for its mechanical requirements. Depending on the engine and its efficiency, the dissipation rate $R_{diss}^{in}$ may be larger than one and the exploitation is then not possible.

The chapter also studies the efficiency of the agro-ethanol industry in the USA. The system currently depends for its requirements on primary energy other than agriculture harvest, and the industries processing these raw energies. Thanks to actual processes which substitute for these auxiliary industries, the agro-ethanol industry can be made self-reliant and rely only on the agriculture harvest.

However, with the current settings of its main operations, can the industry rate $R_{diss}^{in}$ be lower than one, proving the industry viability? And, if it is the case, how it competes in terms of efficiency with the nearly self-reliant petroleum industry for providing transport fuel?

## 5.1. Examples of Self-Reliant Energy Systems

### 5.1.1. Steam Engines and the Thermodynamic Theory

The steam engine conforms to the self-reliant system as described by Fig. 2.2.3 in the section 2.2.2 of the chapter 2, albeit it represents a process rather than an industrial system. The heat value of the coal, or of other fuels, represents the overall input $E_{in}$, while the output of the system $E_{out}$ is the lift work - or electricity today after conversion of the mechanical output -. By energy conservation the dissipation $E_{diss}$ represent the difference between the energies $E_{in}$ and $E_{out}$. The engine efficiency is measured by its dissipation rate, $R_{diss}^{in} = E_{diss}/E_{in}$, or its yield, $Y = E_{out}/E_{in} = 1 - R_{diss}^{in}$.

In this section we use some elements of the thermodynamic theory, the science of conversion of heat into work at its incipience, in order to understand the progresses on steam engine, past and future. The resort to established knowledge to obtain the rate $R_{diss}^{in}$ and the variables influencing it is part of the book methodology. The introduction follows partly the description given by Carnot himself in 1824 [4], as a way to pay tribute to his genial intuitions albeit unachieved. We come back then to the values of the direct rate $R_D$ of various engines given in the section 4.1 of the chapter 4 to convert them into the absolute rate $R_{diss}^{in}$. They are interpreted in light of the theory. Further progresses are indicated.

**Some Elements of Thermodynamics. Engine Operating with Ideal Gas** Carnot's aim was to understand how to obtain the highest amount of work for a fixed input of heat with a fire engine operating between a hot and cold sources each at constant temperature, $T_H$ and $T_C$ respectively.

He made a first generalization by demonstrating that the highest yield should not depend on the working agent of the machine. Otherwise we would have to admit the perpetual motion, or the free creation of work without heat transfer between the two sources.

Because of its inherent imperfections, the cycle of transformations with steam is not able of the highest yield, as seen further down. For his theory Carnot used air and other gases to exploit all the relations and values established experimentally to describe their properties. One of them is the law of the ideal gas, the relation which the absolute pressure $P$, volume $V$ and temperature $T$ of a gas in a known state closely obey:

$$V = n \cdot r \cdot \frac{T}{P}, \tag{5.1.1}$$

where $T$ is in Kelvin, $n$ is the number of moles of the gas in $V$ and $r$ is a constant, $r = 8.31$ J.(K.mol)$^{-1}$.

Besides, for some ideal transformations (reversible expansion and compression in a thermally insulated recipient - or adiabatic transformation) between two states $A$ and $B$ of the gas, the variation of two of its variables is controlled by:

$$\frac{V_A}{V_B} = \left(\frac{T_B}{T_A}\right)^{1/(\gamma-1)}, \tag{5.1.2}$$

where $\gamma$ is the ratio of heat capacities measured in laboratory ($\gamma = 1.4$ for air and about $\gamma = 1.33$ for steam considered as an ideal gas).

The ideal gas and its properties permit to verify Carnot fundamental principle to reach the highest yield. The heat, he also called caloric, should not experience a change of its temperature without doing or receiving a work through a change of the agent volume. The work is maximal when the change can be reversed and the same initial state is obtained again for both engine agent and sources (like in the case of the transformations described by eq. 5.1.2). Hence the ideal yield depends only on the maximum temperature $T_M$ and the minimum one $T_m$ encountered by the agent during its cycle of transformations. In the case of fire engines operating with an ideal gas Carnot assumed perfect heat exchangers such as the temperatures coincides with those of the sources.

He drew a comparison between the caloric and the waterfall driving the wheel of a mill. The work is proportional to the flow - the caloric - and the difference of heights - the difference of extreme temperatures. He assumed, wrongly, that the caloric was conserved like the water flow. However, this point could be distinguished from the principle of reversibility for the maximal yield, as R. Clausius showed with the introduction of the entropy $S$.

To produce work from heat, air enclosed in a cylinder with a movable piston is submitted to a cycle of four ideal transformations:

- The gas receives the heat $E_{in}$ from the hot source at the same temperature $T_H$ of the source ($T_M \equiv T_H$). It starts expanding, producing work.
- Air is then thermally isolated and let again expand. It cools down while producing some work according to Carnot's principle, *i.e.* in a reversible way. The gas reaches the temperature $T_C$ ($T_m \equiv T_C$).
- It is put in contact with the cold source at temperature $T_C$. It transfers the heat $E_{diss}$ to the source at constant temperature, while being compressed.
- Lastly the gas is compressed adiabatically and reversibly to reach the initial temperature $T_H$.

Hence a new cycle with the same transformations can start again. As dilatations occur at temperatures higher than those for compressions, the net production of work is positive and amounts to $E_{out}$.

Because Carnot admitted the heat conservation, *i.e.* $E_{in} \equiv E_{diss}$, he could not derive the relation between the yield of the ideal cycle and its extreme temperatures. Using the thermodynamic properties of air, he derived roughly a work of 3920 $m^3$ of water over 1 m height when the heat of 1 kg of coal, or 7000 kcal, cycles between $T_H = 1000°C$ - at least the temperature of coal combustion - and $T_C = 0°C$ - taking ice as cold source -. When converting both heat and work units into joule unit, the values obtained by Carnot give a yield $Y$ larger than 1.3! Carnot himself expressed some doubts about the heat conservation at the end of his book and in one of the footnotes. In later notes, published long after his death (he prematurely died in 1832), he admitted the heat work equivalence, and even gave a correct coefficient within 20% error interval [4]. But he was not able to reconcile this principle - the first principle in thermodynamics - with the principle given in his book of 1824 - the second one in thermodynamics.

Nevertheless, he was nearly correct when he reckoned that the yield depended on the difference between the temperatures of the hot and cold sources.

In 1850 R. Clausius modified Carnot's theory to make it compatible with the heat work equivalence definitively established by the accurate measurements of J. Joule in 1840's.

To express the Carnot principle of maximal yield at reversibility he introduced the entropy $S$ of a system, besides energy, as the conservative quantity over a cycle of transformations of the system. For a reversible transformation the variation of the entropy is defined as the ratio of the heat transferred to or from the system, to the temperature in Kelvin of transfer (heat transfer and entropy being negative if transfer occurs from the system to a source). If the transformation is irreversible the ratio is lower than the conservative entropy. Integrated over a cycle the ratio is negative. However, when the temperature $T$ of the system is well defined during the transfer and the system is sufficiently simple as to exclude internal irreversibility, its entropy can be still defined as the ratio of heat flow to the temperature $T$. But the whole of the system and its outside is not necessary in equilibrium and the cycle of transformations of the system leads to a growth of the outside entropy, as it will be seen for the steam cycle. In actual heat exchangers the heat transfer $Q$ between a source and the system at temperature $T$ occurs with a small temperature difference $\delta T$ between both such as to assure the transfer, *i.e.* $Q \cdot \delta T > 0$ ($Q$ and $T$ in algebraic notation with the system as the reference). The entropy of the whole rises by $Q \cdot \delta T / T^2 > 0$.

In the case of Carnot engine, we define the engine entropy, $S$, and the source entropies, $S_H$ and $S_C$. As entropy is an additive quantity, the entropy of the whole, $S_{tot}$, is the sum of the part entropies, $S_{tot} = S + S_H + S_C$.

The Carnot cycle of the engine can be represented in a temperature entropy $T - S$ diagram where $T$ is the vertical axis and $S$ the horizontal one (Fig. 5.1.1(a)).

The cycle follows a rectangle like shape. During the heating phase at constant $T_H$, the engine entropy varies from its minimum $S_{min}$ to its maximum $S_{max}$ by the quantity $\Delta S = E_{in}/T_H$ ($\Delta S = S_{max} - S_{min}$). The gas expands then at constant $S$ cooling from the temperature $T_H$ to $T_C$ one. During the subsequent heat transfer at constant $T_C$, the entropy decreases from its maximum $S_{max}$ to its minimum $S_{min}$ by the quantity $-\Delta S = -E_{diss}/T_C$ (all system flows such as $E_{diss}$ are taken positive). Lastly the gas is compressed at constant $S$ and experiences an increase of temperature back to $T_H$.

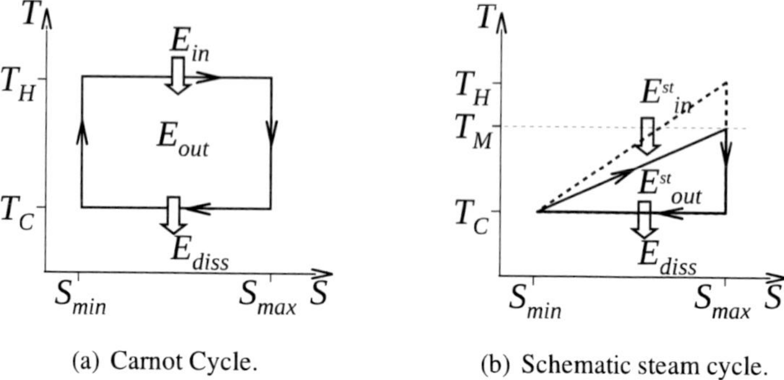


(a) Carnot Cycle. (b) Schematic steam cycle.

Figure 5.1.1. Thermodynamic cycles of the Carnot and steam engines between two heat reservoirs at temperatures $T_H$ and $T_C$ in the temperature entropy diagram.

The area enclosed by the rectangle, $(T_H - T_C) \cdot \Delta S$, corresponds to the net heat disappearance $E_{in} - E_{diss}$, which is also the engine output $E_{out}$.

Over the Carnot cycle the variation of $S$ amounts to $E_{in}/T_H - E_{diss}/T_C$, which must be zero by conservation of $S$.

As a result, the rate $(R^{in}_{diss})_{Carnot}$ for the ideal Carnot engine is:

$$(R^{in}_{diss})_{Carnot} = \frac{E_{diss}}{E_{in}} = \frac{T_C}{T_H}. \tag{5.1.3}$$

Given the temperatures of the sources, the rate is the lowest possible, or the yield the largest, whatever the fluid. For an ideal fire engine working between the hot temperature $T_H = 1\,000°C$, or $1\,273$ K, and the cold one $T_C = 0°C$, or 273.15 K, as suggested by S. Carnot, the rate $R^{in}_{diss}$ is 0.215 or 21.5%, giving a yield $Y$ of 78.5%.

The variation of the entropies of sources are $S_H = -E_{in}/T_H$ and $S_C = E_{diss}/T_C$ respectively. Their sum is thus zero and consequently the overall entropy $S_{tot}$ is also zero. The Carnot cycle is completely reversible.

**Ideal Representation of the Steam Engine.** S. Carnot discarded the steam engine due to the complexity to model it as a result of the change of phase of the water. He also considered rightly this engine not to be able to reach the lowest theoretical rate $(R^{in}_{diss})_{Carnot}$ due to two main irreversibilities.

The steam cycle departs from the Carnot cycle in the stages of compression and heating [54]. In the previous stage, where the fluid transfers its heat to the cold source at temperature $T_m$, the steam condenses to liquid water. Subsequently, the compression of the liquid induces a very low increase of the temperature. Hence the heat transfer from the hot source induces a large rise of the water temperature while little work is produced.

Moreover the highest temperature $T_M$ reached by the steam at the end of the heating stage is lower than the temperature $T_H$ of the hot source.

To draw a comparison with the Carnot cycle, we assume an identical heat flow $E_{diss}$ to the cold source with the transfer occurring at constant temperature $T_m \equiv T_C$. In $T-S$ diagram during this stage the entropy of the engine decreases from its maximum $S_{max}$ to its minimum $S_{min}$ by the quantity $-\Delta S = -E_{diss}/T_C$ (see Fig. 5.1.1(b)).

During the adiabatic expansion of the steam, before cold contact, the entropy $S$ is assumed constant at $S_{max}$, like again in the case of Carnot cycle.

The adiabatic compression stage, however, is supposed to occur at constant $T = T_C$ and constant $S = S_{min}$.

During the stage of heating the engine follows a complex path as the water experiences a change from liquid to gas. The temperature remains constant in the saturated phase where liquid and gas coexist but increases during the pure phases, while $S$ rises in the three phases. Over the overall transformation the entropy $S$ varies from its minimum $S_{min}$ to its maximum $S_{max}$, while the temperature $T$ increases approximately from the lowest temperature $T_C$ to the highest one $T_M$. Assuming that water remains internally in equilibrium with a defined temperature $T$, the heat $E_{in}^{st}$ transferred to the water is:

$$E_{in}^{st} = \int_{S_{min}}^{S_{max}} T(S) \cdot dS.$$

The value depends on the path of the heating stage in the $T-S$ diagram. To simplify we assume a linear relationship between $T$ and $S$:

$$T(S) = T_C + \frac{\Delta T}{\Delta S} \cdot (S - S_{min}),$$

with $\Delta T = T_M - T_C$ (see Fig. 5.1.1(b)). Hence:

$$E_{in}^{st} = (T_C + T_M) \cdot \Delta S/2.$$

By comparison with the Carnot cycle, the output $E^{st}_{out} = E^{st}_{in} - E_{diss}$ is half the value $E_{out}$.

While the variation of $S$ for the engine is still zero over the cycle (by definition), that of the entropies of sources are $\Delta S_H = -E^{st}_{in}/T_H$ and $\Delta S_C = E_{diss}/T_C$ respectively. Their sum is thus not zero:

$$S_H + S_C = \Delta S \cdot \left(1 - \frac{(T_C + T_M)}{2 \cdot T_H}\right),$$

which is positive as $2 \cdot T_H$ is larger than $T_C + T_M$.

Even with $T_M = T_H$ the sum remains positive:

$$S_H + S_C = \frac{\Delta S}{2} \cdot \left(1 - \frac{T_C}{T_H}\right).$$

Consequently the overall entropy $S_{tot}$ is positive. The steam cycle is irreversible as a whole.

The rate $(R^{in}_{diss})_{steam}$ for the simplified steam cycle, but closer to reality than the ideal Carnot engine, is:

$$(R^{in}_{diss})_{steam} = \frac{E_{diss}}{E^{st}_{in}} = 2 \cdot \frac{T_C}{T_C + T_M}. \tag{5.1.4}$$

In modern designs of steam engines the heating stage ends with the water as superheated steam, far from its saturation curve. It permits to increase the temperature $T_M$ towards $T_C$.

Steam reheat operations are also used during the expansion stage. The steam is put in contact with the hot source to increase the heat transfer whereas the temperature difference remains relatively small.

**Comparison of Performances between Actual and Theoretical Steam Engine...** How the theoretical rates from eq. 5.1.3 and 5.1.4 compare with those for some actual steam engines? The question was Carnot's aim at the end of his book when he reported the rates $R_D$ of consumption of engines (see the section 4.1 of the chapter 4), although using a wrong theoretical rate.

To achieve it units of Carnot's data are converted into joule (one $m^3$.m of water lift represents a work of 9 830 $J_m$, while 1 calorie of heat is equivalent to 4.1855 J of work). The resulting rates $R^{in}_{diss}$, and the yields $Y$, for successive designs since the Newcomen engine are reported in Table 5.1.1.

**Table 5.1.1. Characteristics and performances of various actual and theoretical steam engines**

| Engines | $T_M$ in K | $P_M$ in bar | $T_C$ in K | $R^{in}_{diss}$ | $Y$ |
|---|---|---|---|---|---|
| Newcomen | 373 | 1 | 288 | 99.8% | 0.2% |
| Watt 1770 | 373 | 1 | 288 | 99.25% | 0.75% |
| Watt 1824 | 373 | 1 | 288 | 96.5% | 3.5% |
| Wolf 1824 | 408 | 3 | 288 | 96.5-93.5% | 3.5-6.5% |
| Steam cycle | 408 | - | 288 | 83% | 17% |
| Nordjylland 2000 | 855 | 290 | 283 | 53% | 47% |
| Steam cycle | 855 | - | 283 | 50% | 50% |
| Future ? | 1073 | 400 | 283 | $\approx$ 42% | $\approx$ 58% |
| Carnot cycle | 1073 | - | 283 | 26% | 74% |

The table also shows the actual temperatures $T_M$ and $T_m \equiv T_C$ encountered by the water during its cycle, as well as the pressure $P_M$ of the steam at the boiler outlet. The temperatures permit to calculate the rate $(R^{in}_{diss})_{steam}$ for an ideal steam cycle (eq. 5.1.4). For Newcomen and Watt engines S. Carnot noted that the steam came out of the boiler close to its saturation state, *i.e.* the temperature $T_M$ was close to the temperature of saturation of the water at the operating pressure. In the atmospheric engines $T_M$ was close to 100 °C. Even by taking into account the actual temperatures of the steam in eq. 5.1.4, the rates for the engines in Carnot's time remains quite large by comparison with the ideal rate $(R^{in}_{diss})_{steam}$. Imperfections, other than $T_M$ lower than $T_C$ and intrinsic irreversibility of the steam cycle, were losses of the heat of combustion through walls and exhausts, irreversibilities and consequently losses of potential work during heat transfers at the sources and during expansion and compression stages...

The increase of pressure in the Woolf engine permitted a higher saturation point of the steam and so a larger $T_M$, but not that much: for $P_M = 3$ bars $T_M$ was close to 135 °C or 408 K, decreasing the theoretical rate $(R^{in}_{diss})_{steam}$ from 87% - with $T_M = 373$ K for previous engines - to 83%.

The main advantage of the Woolf engine with its high steam pressure and its two different cylinders was to better exploit the temperature difference between the sources.

The industrial production of work by a steam engine has faced and still faces both fundamental and practical limitations to overcome. Given the available cheap sources - fuel combustion in a furnace for a hot one, and water from rivers or sea, or even air, for a cold one - the highest engine yields are reached by making a full profit of the difference of the source temperatures, according to Carnot principle. The mechanical energy is produced during the expansion of the working fluid over this temperature range. A liquid is not adequate due to the poor variation of its temperature and volume over a large change of pressure, as observed for the water compression in the engine. On the other hand, the gas experiences during the adiabatic expansion a large change of its volume given by the relation 5.1.2. For steam approximated to an ideal gas the variation from $V_M$ to $V_C$ is fixed by the relation:

$$\frac{V_C}{V_M} = \left(\frac{T_M}{T_C}\right)^3.$$

The ratio of volumes is thus 2.2 in the case of atmospheric engines, while it is 2.8 for the Woolf engine and 28 for current best turbines (from data in table 5.1.1). Consequently, to limit the size of the installations the volume $V_M$ must be kept the lowest possible. From the law of ideal gas (eq. 5.1.1) the product $V_M \cdot P_M$ is fixed by the temperature $T_M$. The volume can be only diminished by increasing the pressure $P_M$. It corresponds logically to a larger energy density of the steam.

In the Woolf engine the smallest cylinder sustains the high pressure $P_M$ at the start of the expansion stage, for a low volume $V_M$, while the largest one accommodates the high volume $V_C$ of the steam at the end of the expansion. Later the two cylinder role was performed more efficiently by turbines with increasing blade diameters from inlet to outlet.

From 1824 to nowadays, the yield of steam engines has increased steadily, as can be seen in the appendix from the rise of the typical yield of thermal power stations, mainly steam engines: about 15% in 1930's, between 20 and 25% in 1950, 35% in 1975.

The major gains have resulted from the applications of Carnot's theory (use of superheated steam and reheats...), as well as from the improvements to reduce heat losses and to decrease irreversibility during heat transfers (use of heat exchangers operating with low temperature differences, steam turbines instead of cylinders...).

The best current coal fueled power plants using the steam cycle can reach a net efficiency of 47% (LHV basis), such as the Nordjylland 3 station in Denmark. The station produces supercritical steam at 290 bars and $T_M = 582°C$ and operates with reheats during steam expansion, while the cooling occurs at about $T_C = 10°C$ thanks to the cold water of the Baltic sea. The station performance is close to the yield of our ideal steam cycle for its characteristics, $Y_{steam} = 50\%$.

How much further gain can be made on the yield?

Based on our model derived from the thermodynamic laws, the only possibility is to increase the temperature $T_M$ of the steam, which is still far from the temperature of combustion - while the condenser operates already close to the temperatures of available cold sources -, in accordance again with the main recommendation of S. Carnot. The steam pressure $P_M$ to keep the steam volume $V_M$ low is estimated by assuming the same steam energy density as in the Nordjylland 3 station. The enthalpy volumetric content of the steam before expansion is about 290 MJ.m$^{-3}$ (the reference of the enthalpy is fixed by the liquid phase at 0°C).

For a boiler steam at $T_M = 800°C$ the theoretical rate is $(R^{in}_{diss})_{steam} = 0.42$ (or $Y_{steam} = 0.58$), while the pressure $P_M$ is about 400 bars. The ratio of steam volumes $V_M/V_C$ between the start and the end of the adiabatic expansion is close to 55.

New gains are thus possible, provided materials of boilers and turbines can sustain such conditions of pressure, temperature and volume change, and provided heat losses do not rise as a result. However, the increase of the yield won't be more than 15%, not the factor of about 10 obtained since 1824, or 250 since Th. Newcomen.

This section with the presentation of their thermodynamic yields *via* Carnot's data closes the story of the steam engines opened at the beginning of this book.

### 5.1.2. Coal Exploitation in the Loire Basin in 1927

First energy systems had to be self-reliant in the absence of external energies more abundant and cheaper. And even if they had existed poor means of transportation would have limited their export. The most basic source of useful energy, and the dominant one for most part of the History, is the muscular work of humans and animals, which depends on the food supplies (see the basic requirements in the section 3.2.1 of the chapter 3).

Early industrial systems were also self-reliant, even for their equipments, as was the coal exploitation in the basin houiller de la Loire [32]. The coal basin was the oldest exploited in France, since at least the thirteenth century, and the most important one until the middle of the nineteenth. Its last pit closed in 1983, with no much coal left to recover in spite of the difficulty of its exploitation.

An article of 1930 describes the economic and industrial situation of the basin [32]. In 1927 the various producers sold 3 125 kt of cleaned coal out of the 4.0 Mt of raw coal lifted from the mines. Moreover they made 185 kt of products compacted from coal fines (of which 108 kt were used by the local railway company, Paris-Lyon-Marseille). Other thin residues, 617 kt, were submitted to pyrolysis in 320 coke ovens equipped with scrubbers. Hence they produced 542 kt of coke, 15 kt of tar, 3 kt of aromatics and 2.5 kt of ammonia. The remainder, 185 $M.m^3$ of gas - or about 54 kt -, are cleaned and employed for the lightning of towns close to the mines (main coal pits where within Saint-Etienne, the largest city), or as a fuel for nearby industries (glass and cement makers) and for the electric generators of the mines. The generators also used coal filled dusts and fines remaining after all the previous uses.

As a whole the sales of the mines of La Loire amounted in 1927 to nearly 3.9 Mt of products, mostly coal and coke, which corresponded to the net output of the system. Assuming an average high heat value of 29 $GJ_{HHV}.t^{-1}$, due to a content of about 20% of moisture and ash, the output is equivalent to 110 000 $TJ_{HHV}$.

Energy requirements were provided by the electric generators of the mine (55 MW of power in 1927). No data are provided about the type of thermal engines driving the generators. They were probably steam engine based on turbines. From gas of coke and coal dusts they produced 360 $TJ_e$ of electricity for mine requirements. With an assumed net efficiency of about 15% (in LHV basis; see the appendix and the previous section for the base of this hypothesis) the generators consumed about 2400 $TJ_{LHV}$ of fuels. This value is compatible with the use of part of coke-oven gas and of coal fines - less than 0.1 Mt from the mass balance of the system in 1927 -, and with their heat values (about 16 $MJ.m^3$ for the gas).

We can notice that the consumption would be prohibitive with the steam engines of first designs. From the data in Table 5.1.1, the Newcomen type engine would require 180400 $TJ_{LHV}$ of fuels, which is more than the entire production of the basin, the first Watt engines would consume 48 000 $TJ_{LHV}$, and the best of the Watt engines would need about 10 000 $TJ_{LHV}$, making it more reasonable.

The main electrical use was to lift the coal and the excess water from the galleries, often more than 700 m deep. A mass of 5 to 6 tonnes of water per tonne of recovered raw coal had to be pumped because of large infiltrations. Like in the early coal developments, water removal remained a necessity. Hence about 6.5 tonnes of coal and water per extracted tonne of coal had to be lifted, or 26 Mt of material for 1927. For an average depth of 800 m it amounted to 200 $TJ_{meca}$. Once a day 17 600 workers were lifted from the mines. Assuming an average mass of 75 kg/worker and 300 days of work a year the mechanical energy amount to 30 $TJ_{meca}$. The way down for miners and material is not supposed to require energy (it might be possible even to recover part of it). Consequently the lift of coal, water and workers can explain the main part of electricity consumptions. Some was also consumed to produce compressed air for hammer drills.

Due to complex geology with faults and weak rocks, large mechanization was not possible. It implied a total workforce of 28 000 people, of which one third was employed at the surface for the coal cleaning. We may include in the energy consumption the calories of the food necessary for miners to carry out their work. The metabolism of a man requires about 2 700 kcal or 11 MJ of food every day. The requirements for the work at the mine is assumed lower at about 4 MJ per shift. The total calories of food would amount to about 35 TJ, much lower than the fuel consumption of generators.

The transport cost by railway may be also included. We assume a local rate for the transport of order of 1 $MJ_{HHV}.(t.km)^{-1}$, much higher than the current performance of freight transport by train [20] due to the poor efficiency of the steam locomotives at that time. A fraction of 85% of coal was carried over a distance lower than 150 km. We assume an average distance of 100 km. With a HHV of 29 $GJ_{HHV}.t^{-1}$, the coal transport required roughly 0.35% of its HHV. The article reports that the railway bought about 2.5% of the coal output of the mines. However, the trains also carried large part of the manufactured products of the industrialized basin (steel, iron, arms, cycles, glasses, textiles...) between Saint-Etienne and Lyon, as well as the passengers. Besides part of the purchased coal had probably been used for the passenger and freight transport between Paris and Marseille, which route the railway company operated.

Like for the tractor in the agro-ethanol industry we assume the investment consumptions for the mines negligible. Anyway most part of the equipment came from local manufacturers, which bought a small part of the coal, coke and other outputs of the mines to make their products, the largest part being exported.

Consequently, the energy required by the coal industry of la Loire amounted to less than 3% of its output (or coal extracted). The largest part was consumed to pump and lift the water from the deep mines.

This example of a coal basin shows that a self-reliant system is not a pure academic idea. When long distance trade was limited this system was a necessity. It had obviously to be efficient with a large reserve in order to generate an economic growth in the area where it was located. And this was the case.

## 5.2. Case of the Agro-Ethanol Industry in the USA

The current situation of the corn based industry is not that of a self-reliant system.

One of the consequence is the difficulty to measure its efficiency and draw a comparison with other industries such as the petroleum one, as the consumption rate of the industry varies according to the rate definition (details in the section 2.2 of the chapter 2).

To overcome this difficulty the dissipation rate $R^{in}_{diss}$ has been proposed (see the Fig. 2.2.4 of chapter 2), which conforms to the energy conservation. However, the rate can be criticized because the whole system it characterizes includes actually different subsystems. Indeed, besides the ethanol production we have to take into account the petroleum industry to produce diesel, the gas industry to produce ammonia and part of electricity, the coal, nuclear, hydroelectricity and others to produce the rest of electricity, etc. The overall efficiency expressed by $R^{in}_{diss}$ reflects the efficiency of all these systems and not specifically the agro-ethanol industry one.

This section examines how to make the industry self-reliant.

### 5.2.1. A Self-Reliant Agro-Ethanol Industry

So let's imagine some actual processes which permit the agro-ethanol industry in the USA to be self-reliant by resorting only to its agriculture input. When it meets this condition, let's examine whether it is able to generate a large output relative to its input (similarly to the study of the basin de la Loire).

In the new design the processes of the main system remain unchanged. Fig. 5.2.1 shows the direct requirements of the current agro-ethanol industry to process 100 $J_{HHV}$ of corn. The contribution to the output of the co-product - 4.6 $J_D$ - corresponds to the direct energy required at the farm to produce the feed it

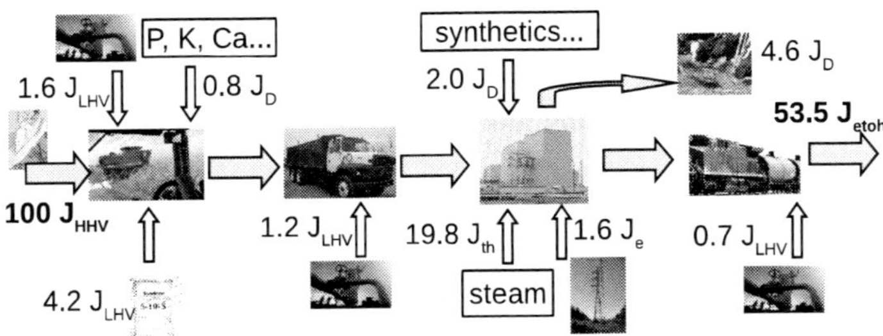


Figure 5.2.1. Direct requirements of the main operations of the current agro-ethanol industry in the USA for a corn input of 100 $J_{HHV}$ [50].

substitutes for. Thereafter it is subtracted from the farm direct consumptions to produce corn.

As the American farms presently do not deliver harvest residues, which could be used as fuels (like in the Brazilian industry with the fiber of the sugar canes) and due to the value of ethanol as fuel, ethanol is considered the substitute for the external energies [50]. The engine maker Scania manufactures a diesel engine operating with ethanol at consumption rates close to the diesel ones in LHV basis. Similarly ethanol can substitute for gas in LHV basis to produce $H_2$ and hence ammonia. Ethanol can also provide heat and electricity at the factory with a cogeneration system working at the overall efficiency of 80% of LHV. Likewise we can find actual or plausible processes using ethanol for all other requirements, even the chemical and material ones, in LHV basis or close.

A more efficient system would fuel the cogeneration process with corn instead of ethanol to avoid the losses of the ethanol production. Corn seems a plausible fuel with a reasonable humidity at 15%, and low ash and tar contents. Obviously corn burning makes this system somewhat plain shocking in face of food shortages in some places of the world. It is not considered here.

The two figures of Fig. 5.2.2 compares the inputs and outputs of the industry with an input of 100 $J_{HHV}$ of corn between current situation and self-reliant one.

A self-reliant American agro-ethanol industry satisfying with part of its outputs all its direct requirements will produce only about 20 $J_{etoh}$ of net output from 100 $J_{HHV}$ of corn, or a rate $R^{in}_{diss} = 80\%$. The ethanol output will decrease from 53.5 J to 15.3 J for 100 $J_{HHV}$ of corn.

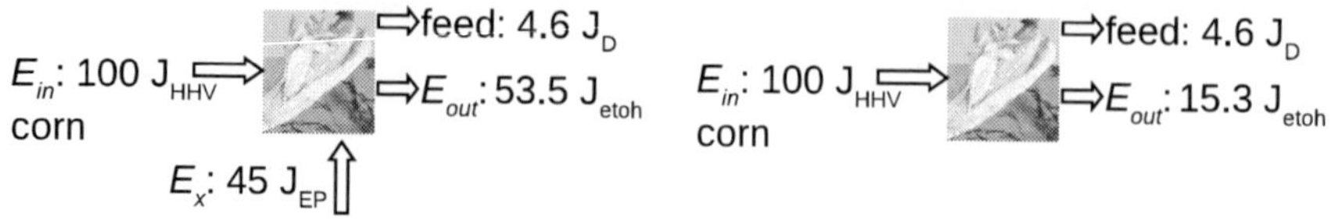

(a) Energy inputs and outputs with a corn input of 100 $J_{HHV}$ in the current industry situation (2007).



(b) Self-reliant system based on existing processes and on substituting ethanol for the direct requirements as shown in Fig. 5.2.1.

Figure 5.2.2. Main energy inputs - corn and external primary energies - and outputs - ethanol and feed - for the production of ethanol in the USA in actual and self-reliant situations.

### 5.2.2. Contribution of the Irrigation

The rate in Fig. 5.2.2 does not take into account the contribution of the irrigation, as its use is not widespread in the corn belt. However, elsewhere like in Nebraska, irrigation is required to compensate for the insufficient rainfalls.

In Nebraska water is pumped from an underground aquifer (see the section 4.3.2 of the chapter 4 for details). Presently either diesel, electrical, or gas engines are used. They provide the mechanical energy $r_{m-pump}$ at the pump outlet to lift and pressurize the water ($r_{m-pump} = 1.0\ \mathrm{MJ}_m.\mathrm{m}_{\mathrm{wp}}^{-3}$ for the average well in Nebraska).

A simple Otto cycle engine using ethanol can supersede the current engines with an overall yield $Y_{Otto}$ of about 30% in LHV basis. The coefficient $\beta_{diss-pump}$ in J ethanol per J of pressure work also comprises the efficiency of the mechanical transmission to the pump - $Y_{tran}$ of about 94% - and the yield of the pump - $Y_{pump}$ of about 75%:

$$\beta_{diss-pump} = Y_{Otto}^{-1} \cdot Y_{tran}^{-1} \cdot Y_{pump}^{-1}\,.$$

Its value is thus $\beta_{diss-pump} = 4.7\ J_{LHV}.J_m^{-1}$.

The contribution of irrigation to the rate $R_{diss}^{in}$ of the industry is then:

$$R_{diss-pump}^{in} = \beta_{diss-pump} \cdot r_{m-pump} \cdot w_{pump}^{in},$$

where the weight $w_{pump}^{in}$ is deduced from weights calculated in the chapter 4:

$$w_{pump}^{in} = w_A^{in} \cdot w_{pump}^A.$$

Its value is $w^{in}_{pump} = 23\ \mathrm{m}^3_{\mathrm{wp}} \cdot \mathrm{GJ}^{-1}_{HHV}$ with a yield $Y_A = 10\ \mathrm{t_A} \cdot (\mathrm{ha} \cdot \mathrm{y})^{-1}$.

The global rate $R^{in}_{diss-pump}$ amounts thus to 11 J of ethanol for 100 J of processed corn.

With the contribution $R^{in}_{diss-pump}$, the overall rate $R^{in}_{diss}$ for the self-reliant industry rises to about 91% - lower than, albeit close to, 100% -. The output available to the end user falls to 9 J for 100 J of processed corn.

### 5.2.3. Comparison with the Petroleum Industry. Resource Limitations

By comparison to produce gasoline the nearly self-reliant petroleum industry has a rate $R^{in}_{diss}$ of about 20% [20].

To qualify these results the agro-ethanol industry relies on a flow resource, which allows a nearly permanent input of plants and so output of ethanol, whereas the crude oil production depends on a limited stock. At some time the oil production must pass a maximum and then decline towards zero. Economic crude oil, with the value of the rate $R^{in}_{diss}$ at 20%, corresponds to a movable oil in a pressurized reservoir rock. Indeed, as seen in the section 3.3 of the chapter 3, hydrocarbon resources in the crust are vast and diversified. However, for the vast majority their heat value would be insufficient to extract them and transform into gasoline or diesel. Currently, to satisfy the demand, a growing part of oil is extracted from tighter rocks with poor extraction yields (less than 5% of the oil in place), and/or with expensive techniques such as water injection and rock heating to make oil move. With such techniques the rate of consumption is increasing. Moreover since 2005 crude oil production has remained limited in spite of the fivefold rise of its price.

On the other hand, a substantial increase of the agro-ethanol production would require an expansion of the planting area from the Corn Belt, where irrigation is small, to less favorable area. According to the US Energy Information Administration the gasoline consumptions in the USA is about 9 millions of barrels per day, or 400 $\mathrm{Mtoe}_{LHV} \cdot \mathrm{y}^{-1}$ (1 toe = 41.9 GJ). The annual corn production for agro-ethanol is about 110 $\mathrm{Mt}_A$, which gives about 23 $\mathrm{Mtoe}_{LHV} \cdot \mathrm{y}^{-1}$ (from the weight $w^{etoh}_A$). To satisfy all the demand corn is already planted in regions too arid to reach the yields of the Corn Belt area, as observed in the Nebraska State.

Moreover, if the industry had to provide internally its requirements to be independent of all fossil energies - including nuclear one -, an annual output of only 6.6 $\text{Mtoe}_{LHV} \cdot \text{y}^{-1}$ would be available, without taking into account the irrigation cost.

## Summary

The present chapter has provided actual examples of self-reliant energy systems, which consumes part of its resource to fuel its processes, from a simple one - the steam engine - to a complex one - the exploitation of the coal basin of la Loire in France. Their efficiency have been estimated by the rate of dissipation $R^{in}_{diss}$ based on the energy conservation (definition in chapter 2).

The steam engine has offered the opportunity to apply established knowledges from both science and industry. In particular the thermodynamic theory has permitted to understand past and future evolutions of the engine efficiency. The theory was funded by S. Carnot in 1824 when he intended to determine the variables driving the efficiency of any fire engine, fundamentally the highest and lowest temperatures encountered by the engine agent during its cycle. Along with practical improvements the theory guided the gains in the rate $R^{in}_{diss}$ of the steam engine, from more than 95% in 1850, to less than 55%. Further reduction requires temperatures and pressures higher than 600°C and 300 bars respectively (800°C and 400 bars for $R^{in}_{diss} = 45\%$), which are limited by material resistance.

The coal mine example has shown the importance of the yield of the steam engine to increase its output. The mine rate $R^{in}_{diss}$ was close to 3% with an engine rate of 85%.

The third example of energy systems in the chapter was the current agro-ethanol industry in the USA, which is not self-reliant. However, it is possible to change its auxiliary processes in order for the industry to draw all its requirements internally, actually from its output of ethanol. The rate of dissipation is particularly high with $R^{in}_{diss}$ =80% (excluding the irrigation consumption).

Even by also taking into account available corn acreage, it seems dubious this resource can substitute for current crude oil production, although the energy is provided by the nature in form of a continuous flow.

# Appendix A

# Appendix: About the Definitions and Conventions on Energy

## Choice of an Energy Unit

A first source of confusion in the energy studies results from the fact that the different communities involved in the various aspects of industrial energies - scientific workers, engineers and economists - do not work with the same unit.

With the establishment of the thermodynamic theory - the science of energy - at around 1850, a unique unit had to be adopted to measure the different forms of energy. Since then the energy unit acknowledged by the Système International d'unités SI has been the joule J (actually joule is a unit derived from the base units like kg, m, s...) [52].

However, many others exist, reminiscence of older units like the calorie, horse-power, Btu... or of practical use in particular area kWh, eV... They each present a fixed factor of conversion into joule, which can be retrieved from scientific handbooks [53] or engineering ones [54]. As a result the conversion between units should not be a difficulty.

Interestingly, both scientific and engineering textbooks do not mention the units employed by the majority of energy agencies, as well as by economists. Thus, to report the statistics of production and consumption, and to present its forecasts, the main energy agency - the International Energy Agency IEA from the Organization for Economic Cooperation - uses the unit toe or ton oil equivalent. Its precise value in joule is 41.868 GJ (the value is an exact multiple of the unit calorie, 1 cal = 4.1868 J, which is probably not a coincidence). The value has been chosen close to the average low heat value LHV of 1 tonne of crude oil. This choice denotes the current economic importance of oil. Due to compositional variation, the actual LHV per tonne depends on the origin of crude oil within some percents.

Coal played previously this economic role and older agencies used the unit tce or ton coal equivalent. Its precise value is 29.3 GJ in J, 6.97 or about 7.0 Gcal in cal, and 0.697 toe [30]. It is noteworthy that S. Carnot also reported a coal of the same LHV per unit of mass to compare the performances of various steam engines (see section 4.1 of the chapter 4). The unit is assumed to coincide with LHV of 1 t of anthracite or 1 t of bituminous coal. However, due to different contents of moisture, ash and volatile matters, LHV of coals, even restricted to bituminous coals, present larger variations than for oil [30, 54].

Through the present book the main unit used is the Si unit, joule. However, the author has found useful to work with the units such as toe or kWh, especially to report the data of consumption or production per year and country produced by the energy agencies (more in the chapter 3 of this book).

### Conventions Adopted to Report Energy Production and Consumption

Data are often presented by the energy agencies with an apparent accuracy of four or more digits. Due to the probabilities of oversight or bad reporting of data, as well as errors of conversion [44], an uncertainty of at least few percent should be factored. It also corresponds to the discrepancies observed when comparing the same data from different sources (after conversion into a same unit).

Beside the problems with gathering the raw data and controlling their accuracy, the difficulties encountered by the employees of the international and national agencies are that of equivalence between the different forms of energy for purpose of comparison and aggregation. Flows of energy are measured by their heat values (after approximate conversions from volume or mass quantities in the case of fuels). However, depending on the forms of energy and on the stage of the production and distribution chain at which their quantities are monitored, the same heat values provide not the same services at the end-user level (like for electricity from primary energies to final user). To solve this dilemma and at the same time to remain close to the principle of energy equivalence, different conventions have been proposed and had existed or exist.

When working with data from different sources (like in the sections 3.1 and

3.2.1), these differences of convention have been taken into account to avoid errors higher than few percents.

Due to their importance, past and present, the conventions examined are that of the statistical office of the United Nation in 1952 [28], the IEA and the Energy Information Administration in USA. The last two agencies were founded in 1974, following the 1973 oil crisis. Conventions of the IEA are adopted by a growing majority of countries.

Primary energy for the IEA are the heat values of the raw resources extracted by the humans and destined mainly for an energy usage (crude oil, gas...) - like in Fig. 3.2.1 of the section 3.2.1 - [44].

The chemical fuels - petroleum, coal, biomass... - are characterized by their LHV (*i.e.* excluding the heat of condensation of the water vapor produced by the combustion), except for the natural gas, which is reported thanks to its high heat value HHV.

The energy from the nucleus of fissile atoms corresponds to the heat of fission produced in the nuclear reactors, obtained from the gross electricity production with an average yield of 33% [1]. Gross production takes into account all the electricity output, even that used internally in auxiliary machines. Besides the primary energy supply, IEA also reports the gross generation of electricity from all generators, whatever the type.

Similarly, primary energy from geothermal heat is obtained from electricity output and a yield of 10%.

On the other hand, energy from resources such as waterfalls, wind, solar rays... are indicated according to the heating value of their gross electricity production.

At the world level the Total Primary Energy Supply corresponds both to global production and consumption, by conservation. At the country level, consumption, or supply, of primary energy differs from the national production depending on the net importation or exportation of the country. The energy kept for the international trades or travels (aviation and marine bunkers, about 3% of world total), and the change of stocks over a year are subtracted from the the annual national supply.

Total final consumption (TFC), shown in Fig. 3.2.2 of the section 3.2.1 for the whole world, is the sum of consumption by the different end-users (from industry to households). For a country it represents the energy consumed by all

[1] the nuclear industry reports for the recent plants higher yields like 36% [30]. The thermal power is deduced from the mass of fission products and the factor of conversion into fission heat

the users within the country boundary. It is obtained from primary energy supply after subtraction of the losses of extraction, conversions and transport. TFC is made of final electricity and processed fuels not used for electricity generation. Hence the nuclear energy disappears from the final consumption due to its use for electricity production. From the world gross electricity generation, 1 723 $Mtep_e$ in 2009, to the final electricity consumption, 1 445 $Mtep_e$, an overall loss of 16% must be taken into account. At the world level, about 6% of the gross generation is consumed for the power plant operations, while about 11% is lost in the transport and distribution networks. The percentage may differ from one country to another (In France and USA the loss is lower than the global average).

A publication from the statistical office of the United Nation in 1952, and extensively used in the the section 3.2.1 of chapter 3, reported for the world and each country the production from *primary sources of energy*, the *gross inland consumption*, the net import or export, and the energy to or from bunkers for international operations [28]. The data correspond to the years 1929, 1937, 1949 and 1950. The *primary sources of energy* refer to the net output of natural and unprocessed sources of energy, whether commercial (coal, petroleum, or hydro power), or non-commercial (wood and wastes). The gross inland consumption takes into account the supply from the international trades, the net quantity drawn from the bunkers and the change of stocks over a year. The definitions of the latter quantities are similar to that used by the IEA, except for electricity production which the UN agency defined as the net production instead of gross one.

The statistical office of the United Nation worked with the unit tce, ton coal equivalent (omitting to indicate the factor of conversion into joule). It converted the raw data - mass, electricity... - about the energy flows into their heat values in tce.

For the masses or volumes of fuels, the agency deduced their *useful calorific values* according to a fixed factor for each category (coal, coke, lignites, crude oil, refined petroleum products, natural gas, coke-oven gas, dry bagasse and peat, fuelwood and wastes). Because the possible variation of LHV or HHV of the different fuels within each category, this convention represents a large oversimplification. Moreover LHV of one t of crude oil is reported as 1.3 tce, or 0.91 toe, instead of 1.0 toe. Actually the equivalent energy of crude oil is defined by the heat value of its refined products obtained from the oil with a loss of mass of 13%.

The heat value of one cubic meter of natural gas is assumed 38.8 $MJ.m^{-3}$, without indication about the conditions of measurement and the grade of the heat value. From the information about gas reported by other agencies [16,44], the conversion factor probably corresponds to the specifications of the gas for pipeline transport in USA, the main producer at that time, i.e. a high heat value of 38.6 $MJ_{HHV}$ per standard cubic meter $m(st)^3$ of gas, standard referring to conditions at temperature 15°C under one atmosphere. It coincides with the convention adopted by IEA for the natural gas.

In a coke oven 1 tonne of coal is assumed to produce 300 $m^3$ of gas, of which heat value is 17.6 $MJ.m^{-3}$ (supposedly $MJ_{HHV}.m(st)^{-3}$).

The electricity production is obtained from the power plant apparent net outputs (or from the plant capacity in absence of more direct data). Two types of generators existed at that time, thermal power plants - mainly coal fueled ones - and hydroelectricity. The equivalent primary energy of the latter is obtained from the product of its output by the average yield of the thermal power plants in the year under consideration. Thus in 1929 and 1937 the yield was 13.7% in LHV basis and in 1950 it rose to 20.5%. This convention, not adopted by IEA, was followed by *l'Observatoire de l'énergie* in France [30] until 2002 and is still applied by the USA Energy Information Administration for its publications. *L'Observatoire de l'énergie* worked with yields of 35.8% in 1975 and 38.5% in 2001. The latter yield was probably that of an oil fueled power plant, as the unit of reference was toe in 2001. The evolution of yields used by the convention is an indication of the progressive improvements of the thermal power plants.

The USA Energy Information Administration (source of Fig. 3.1.2) takes into account all the fuels thanks to their high heat value HHV [16]. They are expressed in the old unit British thermal unit, Btu = 1 054.8 J. Electricity flows correspond to the generator net output.

Primary energy for resources directly delivered in form of electricity such as waterfall and wind are determined from the product of their net electricity production by the average yield of the fossil fuel power plants for the year. The primary energy corresponds to the HHV of fossil fuels displaced by these resources in the electricity production. Thus in 1950 the average electrical yield was 24.5% net/HHV (mostly that of coal fueled plants), 32.8% in 1975 and 34.8% in 2009 (still largely from coal).

The nuclear energy is defined with the same convention as the IEA one.

Because a convention represents some arbitrary choice, a comparison of performance between different forms of energy with data converted according to a convention may be misleading. Primary energy of hydroelectricity is reported in thermal equivalent by USEIA. The international agency reports its gross electricity output, reducing the hydroelectricity weight among primary energy by about one third.

Another choice of primary energy convention would have been the waterfall mechanical energy before generation. Likewise, air kinetic energy flowing through wind turbines could have defined the wind primary energy. The conversion into electricity would have included the average yields of the generators like in the case of more conventional ones. Raw data are the same but the presentation and reported values are different.

# References


[1] E. Fergusson. *Les origines de la machine à vapeur*. In *Histoires de machines*, ouvrage collectif, éditions Belin, 1980 (in French).

[2] A.R. Griffin. *The British coal mining industry: retrospect and prospect*. Moorland, 1977.

[3] J. Payen. Documents relatifs à l'introduction en France de la machine à vapeur de Watt. *Revue d'histoire des sciences et de leurs applications*, 18(3):309–314, 1965 (in French).

[4] S. Carnot (this modern edition contains the original manuscript as well as its critical analysis by R. Fox). *Réflexions sur la puissance motrice du feu, et sur les machines propres à développer cette puissance*. Paris, France: J. Vrin, 1978 (in French).

[5] P. Le Goff. La valeur de l'énergie a-t-elle une base économique, écologique ou technique ? Critère d'optimisation en énergétique industrielle. *Revue d'économie industrielle*, 8(2):68–98, 1979 (in French).

[6] European Fertilizer Manufacturers Association. Understanding nitrogen and its use in agriculture. 2004. 64 pp.

[7] Grécias P. Chimie descriptive organique minérale Math Sup. et Spé PP'Deug A. Paris : Lavoisier, 1988, 447 pp (in French).

[8] M. Appl. The Haber-Bosch Heritage: The Ammonia Production Technology. In *50th Anniversary of IFA technical conference, Sevilla, Spain*, September 25 - 26th 1997.

[9] G. Kongshaug. Energy consumption and greenhouse gas emissions in fertilizer production. In *IFA technical conference, Marrakech, Morocco*, volume 28, page 18, 1998.

[10] European IPPC Bureau at the Institute for Prospective Technological Studies. Reference Document on Best Available Techniques for the Manufacture of Large Volume Inorganic Chemicals - Ammonia, Acids and Fertilizers. EUROPEAN COMMISSION. Aug. 2007.

[11] X. Chavanne and J.P. Frangi. Le rendement énergétique de la production d'éthanol à partir de maïs. *Comptes Rendus Geosciences*, 340(5):263–287, 2008 (In French).

[12] X. Chavanne and J.P. Frangi. Comparison of the energy efficiency to produce agroethanol between various industries and processes: the farm stage. *Biomass and Bioenergy, submission process since* March 2010.

[13] International Energy Agency. Tracking Industrial Energy Efficiency and CO2 Emissions. *Published by the Joint Research Centre of the European Commission*. Pages: 621. July 2007.

[14] R. Remus et al. Best Available Techniques Reference Document for Iron and Steel Production. *Published by : OECD Publishing*. Pages: 324. 2013.

[15] M. Pénin. *Comptabilité nationale*. Encyclopedia Universalis, France, 2003 (In French).

[16] USA Energy Information Administration. Annual energy review, data. http://www.eia.gov/totalenergy/data/annual/index.cfm, October 2011.

[17] US census bureau. Americanfact finder. http://factfinder2.census.gov/faces/nav/jsf/ pages/searchresults.xhtml?, retrieved end of january 2012.

[18] USA Energy Information Administration. Natural gas, data, prices. http://205.254 .135.24/naturalgas/data.cfm#prices, 2011.

[19] USA Energy Information Administration. petroleum, data, refining. http://205.254.135.24/petroleum/data.cfm#refining, October 2011.

[20] X. Chavanne and J.P. Frangi. Comparison of the energy efficiency to produce agroethanol between various industries and processes: the transport stage. *Biomass and Bioenergy*, 35(9):4075–4091, October 2011.

[21] W.Y. Huang. Impact of rising natural gas prices on u.s. ammonia supply. wrs-0702. Technical report, U.S. Department of Agriculture, Economic Research Service, August 2007.

[22] T. Woody. Solar Power's Dark Side. New York Times, page B1 of the New York edition. May 29, 2013.

[23] Business Insights Ltd. The solar cell production global market outlook, June 2011.

[24] G. Hagedorn, E. Hellriegel, Umwelrelevante Masseneinträge bei der Herstellung verschiedener Solarzellentypen; etc. -Endbericht -Teil I: Konventionelle Verfahren, Forschungstelle für Energiewirtschaft: München, Germany, (1992).

[25] K. Bradsher. Strategy of Solar Dominance Now Poses a Threat to China. New York Times, page B1 of the New York edition. October 5, 2012.

[26] EPIA. Global-Market-Outlook-for-Photovoltaics-until-2015. May, 2012.

[27] Ø. Ulleberg, T. Nakken, and A. Eté. The wind/hydrogen demonstration system at Utsira in norway: Evaluation of system performance using operational data and updated hydrogen energy system modeling tools. *International Journal of Hydrogen Energy*, 35(5):1841–1852, 2010.

[28] Statistical office of the United Nations. World energy supplies in selected years, 1929-1950. New York, 1952.

[29] International Energy Agency. Key world energy statistics. Paris, France, 2011.

[30] CEA. *Memento Sur l'Energie - Energy Handbook*. Edition 2011. France, 2011.

[31] E. Fergusson. *Les origines de la machine à vapeur*. In *Histoires de machines*, ouvrage collectif, éditions Belin, 1982 (in French).

[32] M. Perrin. Le bassin houiller de la loire. In *Annales de Géographie*, volume 39, pages 359–375. Société de géographie, 1930 (In French). See also http://fr.wikipedia.org/wiki/Bassin_houiller_de_la_Loire.

[33] E. Leroy Ladurie. *Histoire humaine et comparée du climat: Canicules et glaciers XIIIe-XVIIIe siècles*. Tome 1. Paris, Fayard 2004.

[34] S.C. Davis, S.W. Diegel, and R.G. Boundy. Transportation energy data book: Edition 30. Technical report, Oak Ridge National Laboratory. Center for Transportation Analysis and United States. Dept. of Energy. Office of Energy Efficiency and Renewable Energy, 2011.

[35] X. Chavanne and J.P. Frangi. De la détermination du rendement des filières énergétiques. *Comptes Rendus Geosciences*, 339(8):519–535, 2007 (In French).

[36] R. Delmas, G. Megie, and V.H. Peuch. *Physique et chimie de l'atmosphère*. Coll. Echelles, Belin, France, 2005, (In French).

[37] M.K. Hubbert. Energy resources of the earth. *Sci. Am.*, 225:31–40, September 1971.

[38] B. Durand. *Definition and quantitative importance of kerogen*. In *Kerogen: insoluble organic matter from sedimentary rocks*, Editions technip, 1980.

[39] J.M. Hunt. Distribution of Carbon in Crust of Earth. AAPG bulletin 56(11):2273–2277, 1972.

[40] J.M. Hunt. Distribution of Carbon as Hydrocarbons and Asphaltic Compounds in Sedimentary Rocks. AAPG Bulletin 61(1), 100–104, 1977.

[41] R.C. Weast, M.J. Astle, and W.H. Beyer. *CRC handbook of chemistry and physics*, volume 69. CRC press Boca Raton, FL, 1988.

[42] Rössing Uranium Mine. 2010 report to stakeholders. http://www.rossing.com, 2011.

[43] Natural Resources Canada and Mining Association of Canada. Benchmarking the energy consumption of canadian open-pit mines. Technical report, Canada, 2005.

[44] IEA, OCDE and Eurostat. Energy statistics manual. IEA publications, Paris France, September 2004.

[45] J.L. Sullivan, C.E. Clark, J. Han and M. Wang. Life-cycle analysis results of geothermal systems in comparison to other power systems. Technical report, Argonne National Laboratory (ANL), 2010.

[46] M.K. Hubbert. Nuclear energy and the fossil fuel. *Drilling and production practice*, 1956.

[47] J. Barton and C. Guillemet. *Le verre, science et technologie*. EDP sciences, 2005 (In French).

[48] M.L. Socolof, J.G. Overly, and J.R. Geibig. Environmental life-cycle impacts of CRT and LCD desktop computer displays. *Journal of Cleaner Production*, 13(13-14):1281–1294, 2005.

[49] SAMSUNG ELECTRONICS. 2008-2009 sustainability report, 2009.

[50] X. Chavanne and J.P. Frangi. Comparison of the energy efficiency to produce agroethanol between various industries and processes: Synthesis. *Biomass and Bioenergy*, 35(7):2737–2754, July 2011.

[51] X. Chavanne and J.P. Frangi. Comparison of the energy efficiency to produce agroethanol between various industries and processes: The factory stage. *Biomass and Bioenergy, submission process since* March 2010.

[52] Le Système international d'unités (SI), 8e édition/The International System of Units (SI), 8th edition, ed. by Bureau international des poids et mesures, STEDI Media, Paris 2006.

[53] R. C. Weast et al. *Handbook of physics and chemistry*. CRC Press, Boca Raton. 1986.

[54] F. Massard. *Aide-mémoire du thermicien*. Ed Elsevier : Paris. 526 pp. 1997 (In French).

# Index

## A



## B



## C

## D

## E

## F



## G



## H

## I



## J



## K



## L

## M



## N



## O



## P

## Q



## R

## S



## T

## U



## V



## W



## Y